# Fun with Everyday Math

by

Daryl Vriesenga

Cover Artist
Vickie Lane

2 1

Inside Illustrations by
Chris Olsen

*Publishers*
Instructional Fair • TS Denison
Grand Rapids, Michigan 49544

## Permission to Reproduce



## About the Author

**Daryl Vriesenga** holds a master's degree from Michigan State University and has been a teacher at the elementary level for over 20 years. This year he assumed the responsibilities of principal at the school where he teaches. He is also the author of Instructional Fair • TS Denison's *Science Enrichment Series, Science Fair Projects,* and *Earth Science* as well as other popular Instructional Fair • TS Denison products.

## Credits

Author: Daryl Vriesenga
Cover Artist: Vickie Lane
Inside Illustrations: Chris Olsen
Project Director/Editor: Linda Kimble
Editors: Elizabeth Flikkema, Eunice Kuiper
Typesetting/Layout: Pat Geasler
Cover Photos: © Image Club, © Digital Stock © Corel, © Foto Rom, and © Instructional Fair • TS Denison

Standard Book Number: 1-56822-608-X
*Fun with Everyday Math*

2400 Turner Avenue NW
Grand Rapids, Michigan 49544

# Table of Contents

# Name That Number

Use < and > to compare each pair of numbers.

1. 3,642 ________ 4,693
2. 6,418 ________ 6,481
3. 21,009 ________ 20,999
4. 13,391 ________ 13,399
5. 2,791 ________ 2,971
6. 781,444 ________ 781,409
7. 1,109,322 ________ 1,108,999
8. 3,330,001 ________ 3,330,987

Circle the number that is . . .

| | | | | |
|---|---|---|---|---|
| 9. | less than 937. | 937 | 939 | 934 |
| 10. | greater than 4,671. | 4,641 | 4,599 | 4,700 |
| 11. | less than 21,413. | 13,991 | 21,415 | 21,600 |
| 12. | less than 2,333 and greater than 2,105. | 2,567 | 2,222 | 2,101 |

| State | Area in Square Miles |
| --- | --- |
| Illinois | 57,918 |
| Indiana | 36,420 |
| Iowa | 56,276 |
| Michigan | 96,705 |
| Minnesota | 86,943 |
| Ohio | 44,828 |

Round off the area of each of the states. Complete the chart.

| State | Round to the nearest . . . | | |
| --- | --- | --- | --- |
| | ten-thousand | thousand | hundred |
| Illinois | | | |
| Indiana | | | |
| Iowa | | | |
| Michigan | | | |
| Minnesota | | | |
| Ohio | | | |

# What's Missing?

Complete each of these number patterns.

1. 36 ___ ___ 24 20 16 ___ ___ 4 ___
2. 21 ___ ___ 42 ___ 56 63 ___ 77 ___
3. 24 ___ 40 ___ ___ 64 72 80 ___ ___
4. ___ 45 ___ 63 72 81 ___ ___ 108 ___
5. 2 7 ___ ___ 22 ___ 32 ___ 42 ___
6. 41 46 ___ ___ 61 ___ 71 ___ 81 ___
7. 13 ___ 27 34 ___ 48 55 ___ ___ 76
8. 120 ___ 116 114 ___ 110 ___ ___ 104 102
9. 240 235 ___ 225 ___ ___ 210 205
10. 103 106 ___ ___ 115 118 ___ 124 ___ ___

Study the numbers and complete the pattern.

99 118 137 _____ _____ _____ _____

_____ _____ _____ 109 105 101 _____

400 _____ 412 418 _____ _____ _____

842 836 _____ 824 _____ _____ _____

224 221 _____ 215 _____ _____ _____

207 _____ 221 _____ 235 _____ 249 _____

1,025 _____ 975 _____ 925 _____ 875 _____

110 _____ 70 50 30 _____ _____ _____

99 _____ 117 126 _____ _____ 153 _____

1,002 1,004 1,008 1,016 _____ _____ _____

# Just Write It!

| Simple number | | Expanded number |
|---|---|---|
| 41,673 | = | 40,000 + 1,000 + 600 + 70 + 3 |

Write each expanded number in its simple form.

1. 30,000 + 5,000 + 900 + 40 + 7 ______________
2. 60,000 + 800 + 50 + 6 ______________
3. 5,000 + 40 + 3 ______________
4. 80,000 + 3,000 + 80 + 6 ______________
5. 10,000 + 800 + 50 ______________

Write each simple number in its expanded form.

6. 32,463 ______________________________
7. 5,245 ______________________________
8. 3,006 ______________________________
9. 40,197 ______________________________
10. 9,100 ______________________________

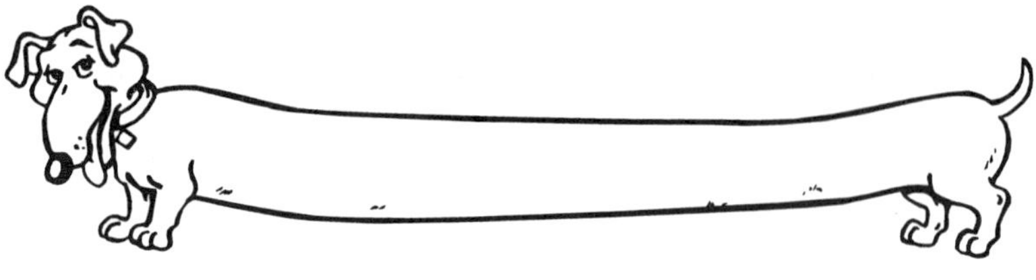

| Numeral | | Word Form |
|---|---|---|
| 26,874 | = | twenty-six thousand, eight hundred seventy-four |

Write the numeral for each word form.

1. two thousand, five hundred sixty-one ____________
2. twenty-three thousand, six hundred ninety-four ____________
3. seventy-six thousand, eighty-six ____________
4. four hundred thirteen thousand, one hundred seven ____________
5. nine hunded forty-one thousand, one hundred ten ____________

Write each simple number in word form.

6. 1,437

____________________________________

7. 21,900

____________________________________

8. 80,063

____________________________________

9. 10,013

____________________________________

10. 123,004

____________________________________

 

# Estimation Game

Before you add, estimate the answer by rounding each number to its highest place value. After adding, compare the estimated sum with the actual sum. If your sum and estimation are not close, recheck your addition.

| Actual | Estimated |
|---|---|
| 392 | 400 |
| + 421 | + 400 |
| **813** | **800** |

1. 296 + 342 = ____ Estimate: ☐☐☐ + ☐☐☐ = ____

2. 4,726 + 3,121 = ____ Estimate: ☐☐☐☐ + ☐☐☐☐ = ____

3. 782 + 413 = ____ Estimate: ☐☐☐ + ☐☐☐ = ____

4. 2,827 + 8,293 = ____ Estimate: ☐☐☐☐ + ☐☐☐☐ = ____

5. 29 + 36 + 78 = ____ Estimate: ☐☐ + ☐☐ + ☐☐ = ____

6. 85 + 97 + 23 = ____ Estimate: ☐☐ + ☐☐☐ + ☐☐ = ____

Round each number to the nearest ten or hundred. Then multiply.

1. 296 x 18 = ______
2. 961 x 53 = ______
3. 423 x 72 = ______
4. 83 x 94 = ______
5. 623 x 595 = ______
6. 422 x 68 = ______
7. 32 x 38 = ______
8. 601 x 8 = ______

9. It is a 323 mile round trip between Grand Rapids, Michigan, and Chicago. The Amtrak train makes the round trip 62 times each month. Approximately how many miles does it travel in one month?

______

10. If the train averages 135 passengers on each round trip, and the train makes 62 round trips each month, approximately how many passengers make the trip each month?

______

# Travel Yellowstone

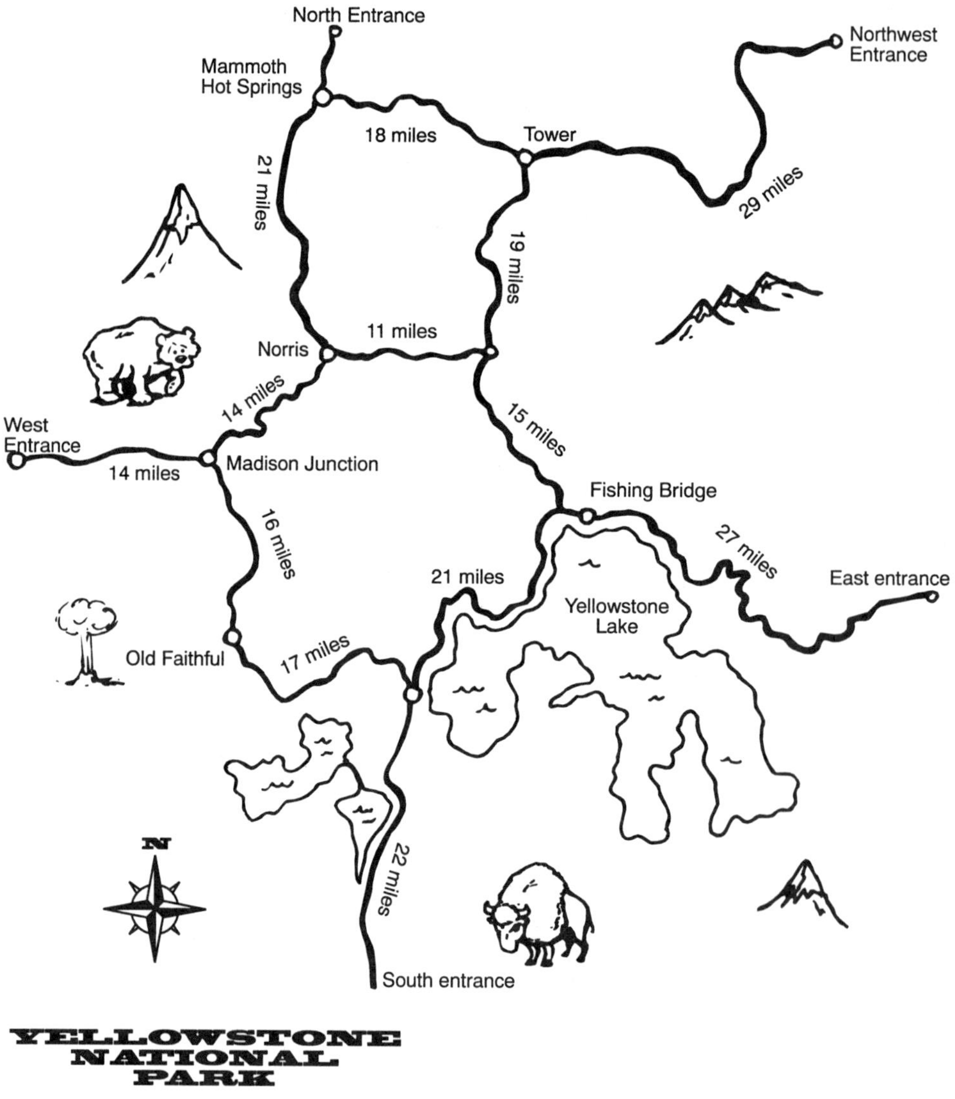

The Williams family is taking a family vacation in Yellowstone National Park. They will drive to see many attractions in this large park.

Refer to the road map to calculate the shortest mileage they can travel between the locations listed.

First estimate the mileage between the two places named. Then calculate the actual mileage.

| East Entrance and South Entrance | Madison Junction and Tower |
|---|---|
| Estimate Actual | Estimate Actual |

| West Entrance and East Entrance | Northwest Entrance and South Entrance |
|---|---|
| Estimate Actual | Estimate Actual |

| West Entrance and Fishing Bridge | Old Faithful and Mammoth Hot Springs |
|---|---|
| Estimate Actual | Estimate Actual |

# Subtraction Action

Rename the tens and show ten more ones. Then subtract.

**Example:**

```
   5 13
  4 6 3   (6 and 3 crossed out)
– 2 4 6
-------
  2 1 7
```

Rename the hundreds and show ten more tens. Then subtract.

**Example:**

```
 6 12
 7 2 9   (7 and 2 crossed out)
– 4 5 2
-------
  2 7 7
```

Find the difference.

1. 652 − 239 = ____
2. 847 − 493 = ____
3. 421 − 291 = ____
4. 4,683 − 2,591 = ____
5. 9,364 − 4,272 = ____
6. 9,601 − 7,281 = ____
7. 5,381 − 2,273 = ____
8. 7,736 − 5,728 = ____
9. 9,223 − 7,107 = ____

Regroup and subtract to find the difference.

**Examples:**

| | |
|---|---|
| 3 13 | 2 13 13 |
| 343 | 343 |
| – 165 | – 165 |
| 8 | 178 |

1. 920 – 729

2. 856 – 298

3. 621 – 187

4. 867 – 268

5. 371 – 189

6. 863 – 378

7. 881 – 494

8. 641 – 258

9. 752 – 268

# Larry's Lawn and Garden Center

grass seed
$2.49 lb.

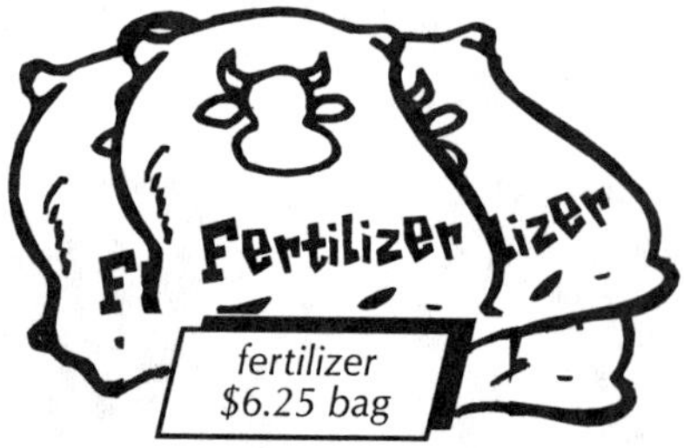

fertilizer
$6.25 bag

fence posts
$5.76 each

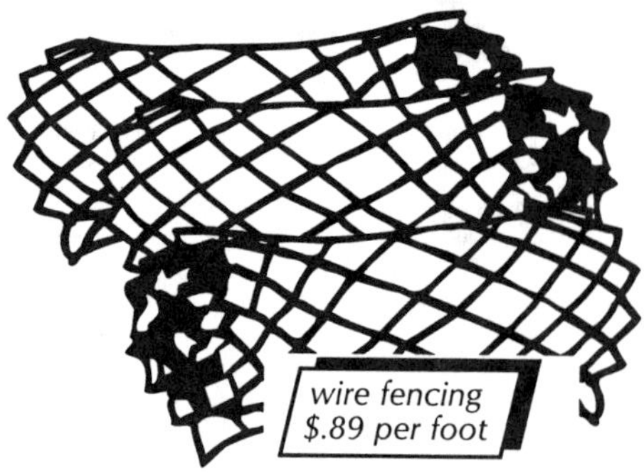

wire fencing
$.89 per foot

cement
$3.47 bag

flowers
$6.79 per flat

tomato plants
$1.39 each

rose bushes
$9.39 each

tomato stakes
$0.79 each

It's a warm, sunny Saturday in spring, and Larry's Lawn and Garden Center is a busy place. What will it cost each family to purchase its list of supplies for their lawns and gardens? Show your work. Remember you're computing money and label your answers appropriately. (There is no tax on the items.)

1. The Williams family needs 15 pounds of grass seed, two bags of fertilizer, and six rose bushes.

2. The Johnson family wants to build a fence. They need 75 feet of wire fencing, eight fence posts, and six bags of cement.

3. The Ruiz family wants to plant a garden. They need eight tomato plants, eight tomato stakes, six flats of flowers, and two bags of fertilizer.

# Measure Up!

Mark each measurement on the inch ruler. The first one is done for you.

**A.** 1¾ inches

**B.** 5⅜ inches

**C.** 4⅞ inches

**D.** 3¼ inches

**E.** 2½ inches

Mark each measurement on the centimeter ruler.

**F.** 6 centimeters

**G.** 4½ centimeters

**H.** ½ centimeter

**I.** 25 millimeters

**J.** 72 millimeters

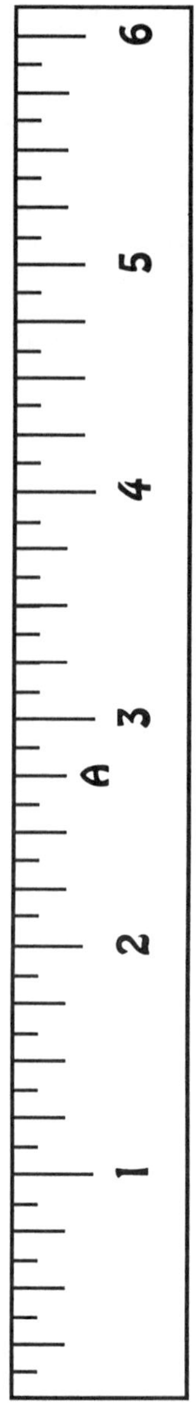

Circle the unit of measurement that you would use to measure each object.

Write the matching letters in the spaces at the bottom of the page to solve the riddle.

**1.**

S kilometers
R meters
W centimeters

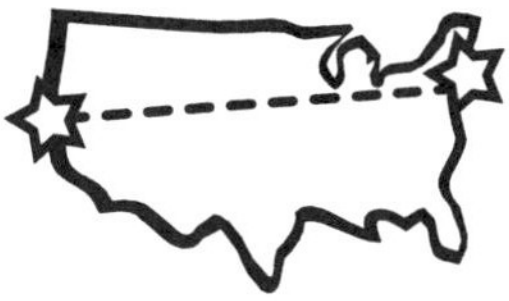

**2.**

R kilometers
B meters
C centimeters

**3.**

U meters
A millimeters
B centimeters

**4.**

A miles
E feet
U inches

**5.**

S inches
R miles
T yards

**6.**

L inches
K feet
S yards

**"What kind of water should you drink if you are sick?"**

| ___ | ___ | ___ | ___ | | ___ | ___ | ___ | ___ | ___ |
|---|---|---|---|---|---|---|---|---|---|
| 1 | 4 | 6 | 6 | | 1 | 3 | 5 | 4 | 2 |

# How Big Is It?

Determine the perimeter of each of these figures.

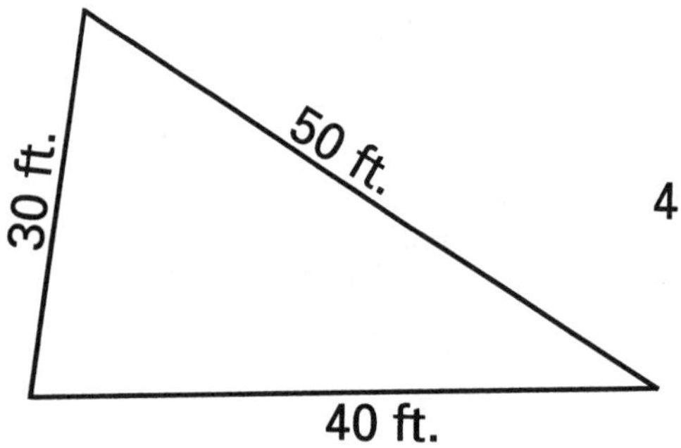


_______ feet

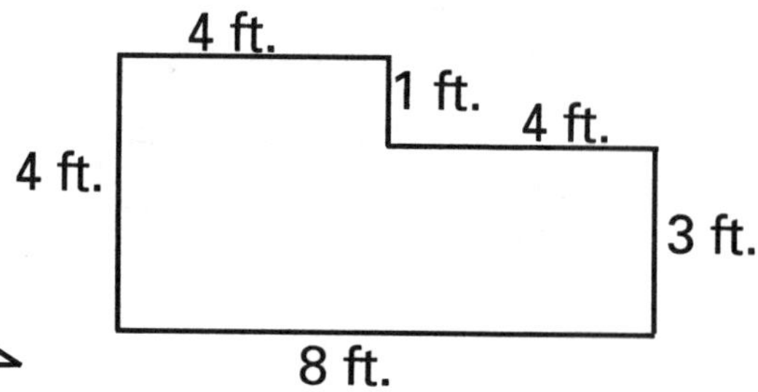


_______ feet

1. Mr. Anderson's garden is in the shape of a 65' x 25' rectangle. How many feet of fencing would it take to enclose the garden?

2. Coach Fritz told his team to run one time around the football field. The field is 360 feet long and 160 feet wide. How far do they have to run?

Find the area of each figure in square feet.

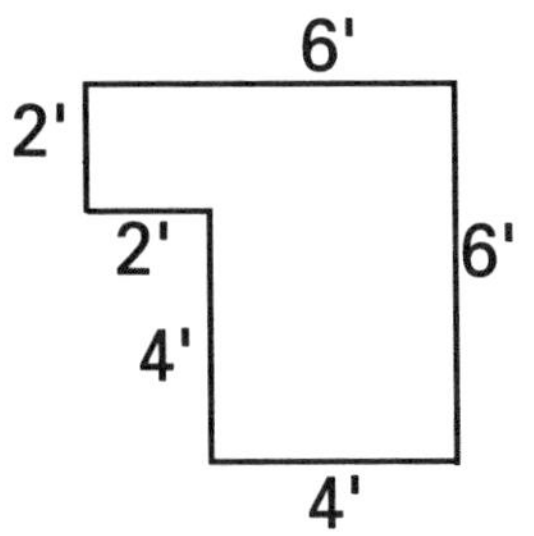


____ square feet

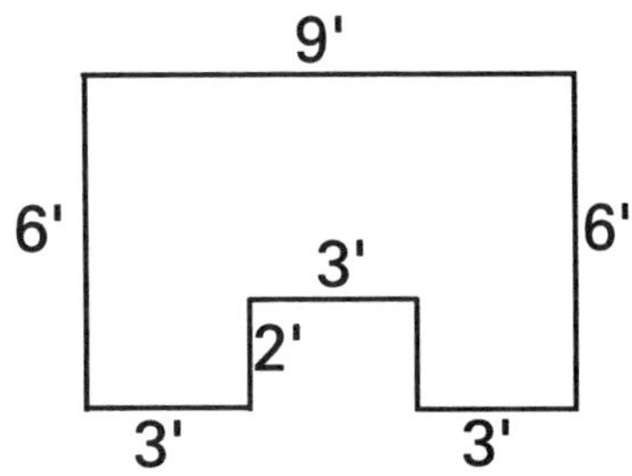


____ square feet

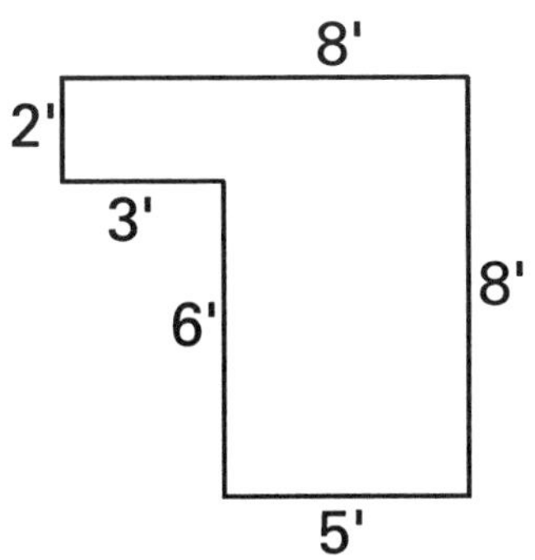


____ square feet

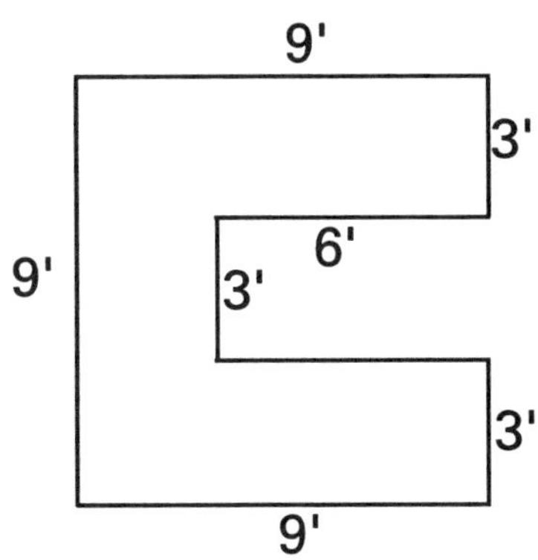


____ square feet

# Fraction Action

Write the fractional parts on each figure in simplest terms.

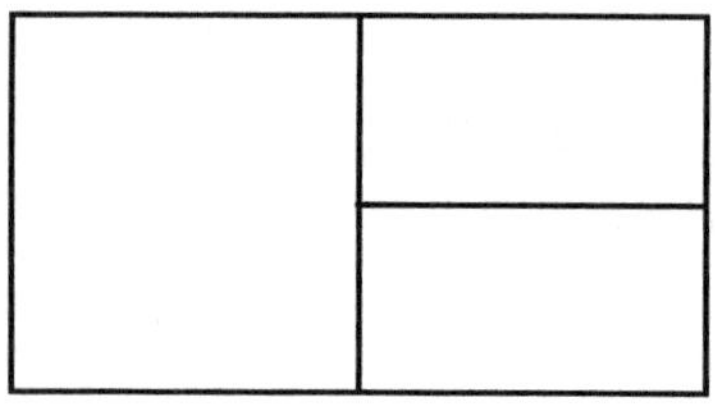

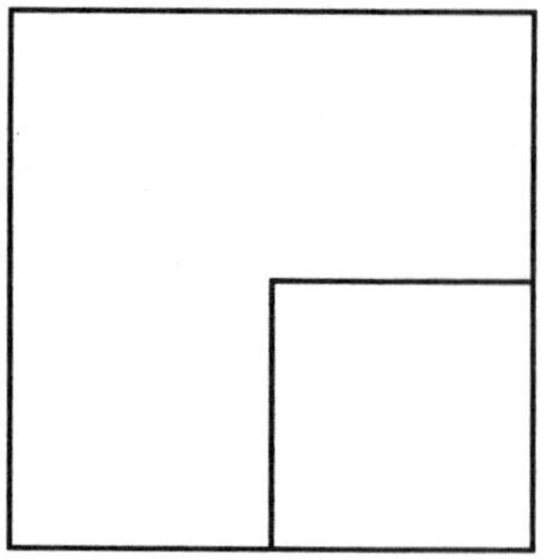

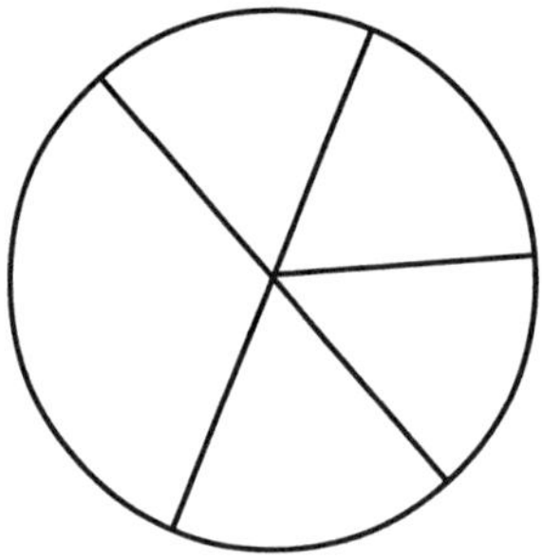

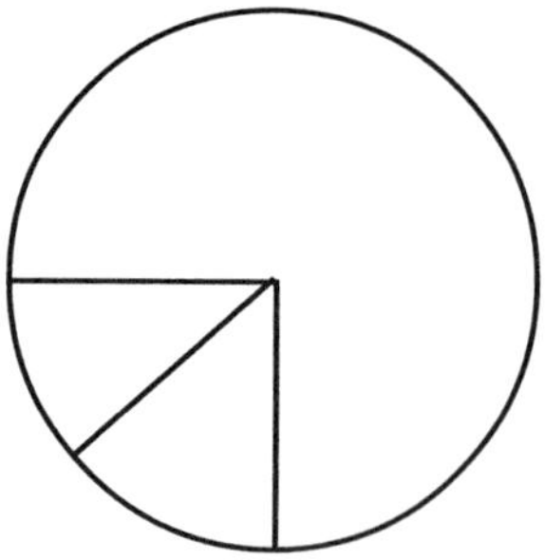

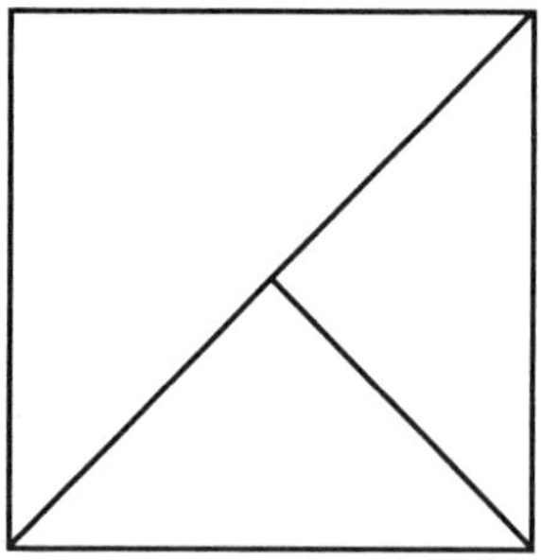

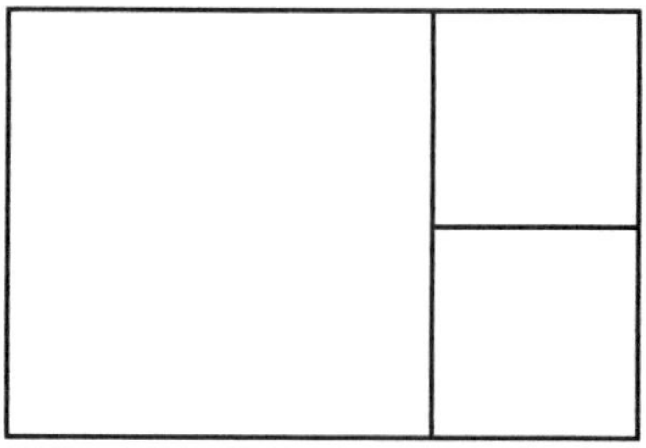

**"If buttercups are yellow, what color are hiccups?"**

____ ____ ____ ____ ____ ____

Write each fraction in simplest form.

$\frac{3}{6} =$ $\frac{9}{21} =$ $\frac{6}{9} =$

$\frac{5}{20} =$ $\frac{16}{20} =$ $\frac{10}{16} =$

Cross out each simplified fraction and the letter above it. The remaining letters will solve the riddle.

| C | B | R | U | L | R | G | P | N | L | E | R |
|---|---|---|---|---|---|---|---|---|---|---|---|
| $\frac{1}{2}$ | $\frac{3}{6}$ | $\frac{2}{3}$ | $\frac{6}{10}$ | $\frac{3}{7}$ | $\frac{4}{12}$ | $\frac{1}{4}$ | $\frac{6}{9}$ | $\frac{4}{5}$ | $\frac{2}{4}$ | $\frac{4}{8}$ | $\frac{5}{8}$ |

# Another Name

Circle the fraction in each row that is not equivalent to the other four.

1. $\frac{1}{2}$ $\frac{2}{4}$ $\frac{6}{16}$ $\frac{3}{6}$ $\frac{16}{32}$

2. $\frac{5}{15}$ $\frac{2}{3}$ $\frac{8}{12}$ $\frac{6}{9}$ $\frac{16}{24}$

3. $\frac{9}{12}$ $\frac{6}{8}$ $\frac{3}{4}$ $\frac{15}{20}$ $\frac{20}{25}$

4. $\frac{4}{16}$ $\frac{4}{12}$ $\frac{5}{20}$ $\frac{1}{4}$ $\frac{2}{8}$

5. $\frac{1}{5}$ $\frac{6}{60}$ $\frac{2}{10}$ $\frac{5}{25}$ $\frac{10}{50}$

Write three equivalent fractions for each given fraction.

1. $\frac{1}{3}$

2. $\frac{3}{5}$

3. $\frac{1}{10}$

Find the equivalent fraction.

1. $\frac{1}{2} = \frac{\_\_}{4}$

2. $\frac{2}{3} = \frac{6}{\_\_}$

3. $\frac{3}{4} = \frac{\_\_}{16}$

4. $\frac{2}{8} = \frac{\_\_}{24}$

5. $\frac{\_\_}{5} = \frac{3}{15}$

6. $\frac{3}{\_\_} = \frac{6}{16}$

7. $\frac{2}{\_\_} = \frac{8}{40}$

8. $\frac{1}{\_\_} = \frac{5}{20}$

# Add 'Em Up

Write the addition problem represented by each pair of fractions. Write the sum in simplest terms.

1. 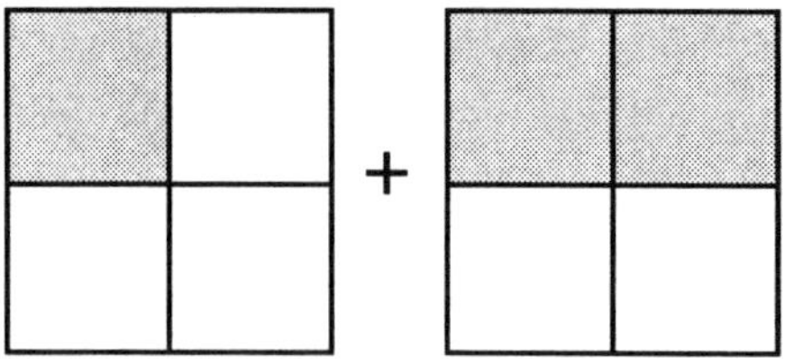

2. 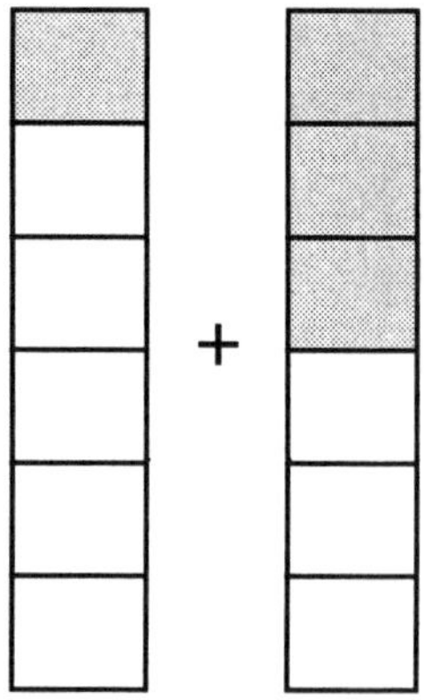

3. 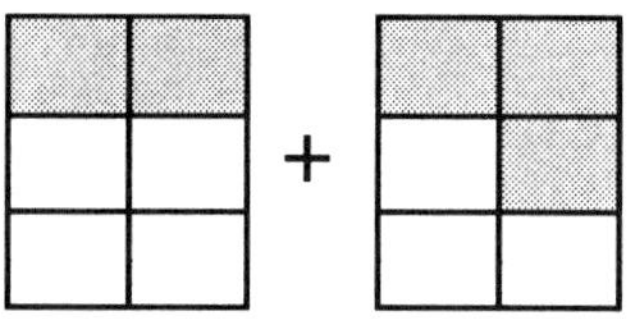

4. 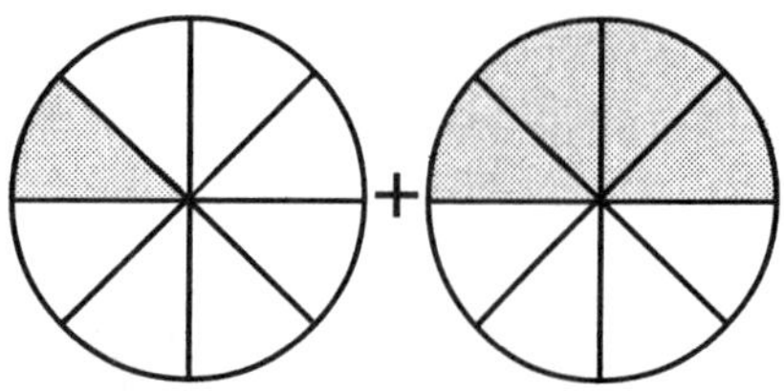

5. 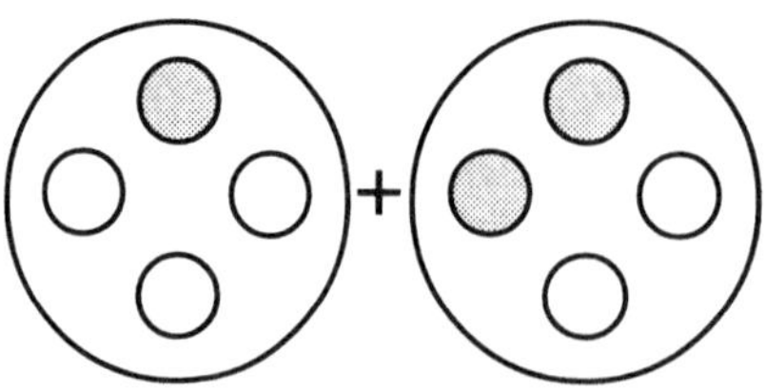

6. 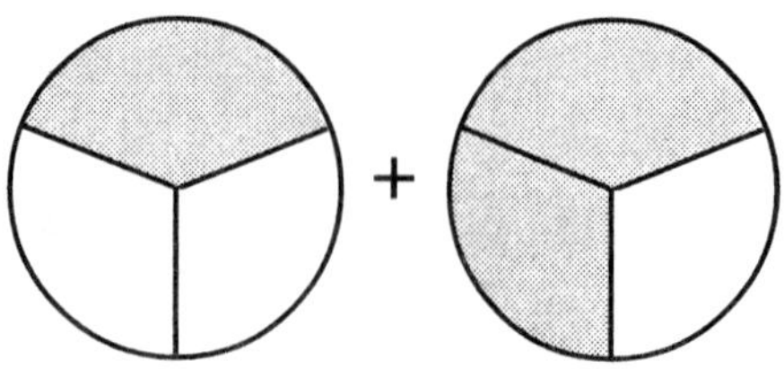

Find the sums. Write the sums in simplest terms and rename improper fractions as mixed numbers.

1. $\frac{7}{8} + \frac{4}{8} =$

2. $\frac{3}{16} + \frac{7}{16} =$

3. $\frac{1}{5} + \frac{3}{5} =$

4. $\frac{4}{8} + \frac{5}{8} =$

5. $\frac{5}{9} + \frac{2}{9} =$

6. $\frac{3}{4} + \frac{3}{4} =$

7. $\frac{2}{6} + \frac{1}{6} =$

8. $\frac{6}{10} + \frac{8}{10} =$

9. $\frac{3}{8} + \frac{1}{8} =$

10. $\frac{7}{8} + \frac{3}{8} =$

# Dancing Denominators

Solve by finding the least common denominator. Write the sum or difference in simplest terms.

**Example:**

$$\begin{array}{rcr} \frac{3}{4} & = & \frac{6}{8} \\ -\frac{1}{8} & = & -\frac{1}{8} \\ \hline & & \frac{5}{8} \end{array}$$

1. $$\begin{array}{rcr} \frac{1}{3} & = & \_\_ \\ +\frac{3}{6} & = & +\_\_ \\ \hline & & \end{array}$$

2. $$\begin{array}{rcr} \frac{7}{16} & = & \_\_ \\ -\frac{2}{8} & = & -\_\_ \\ \hline & & \end{array}$$

3. $$\begin{array}{rcr} \frac{1}{2} & = & \_\_ \\ -\frac{1}{8} & = & -\_\_ \\ \hline & & \end{array}$$

4. $$\begin{array}{rcr} \frac{2}{5} & = & \_\_ \\ +\frac{3}{10} & = & +\_\_ \\ \hline & & \end{array}$$

5. $$\begin{array}{rcr} \frac{1}{8} & = & \_\_ \\ +\frac{3}{16} & = & +\_\_ \\ \hline & & \end{array}$$

Solve by finding the least common denominator. Write improper fractions as mixed numbers in their lowest terms.

**Example:**

$$\begin{array}{rcr} \frac{4}{5} & = & \frac{12}{15} \\ -\ \frac{1}{3} & = & -\ \frac{5}{15} \\ \hline & & \frac{7}{15} \end{array}$$

1. $$\begin{array}{rcl} \frac{3}{4} & = & \_\_ \\ +\ \frac{1}{3} & = & +\ \_\_ \\ \hline \end{array}$$

2. $$\begin{array}{rcl} \frac{1}{3} & = & \_\_ \\ -\ \frac{2}{8} & = & -\ \_\_ \\ \hline \end{array}$$

3. $$\begin{array}{rcl} \frac{1}{6} & = & \_\_ \\ +\ \frac{3}{8} & = & +\ \_\_ \\ \hline \end{array}$$

4. $$\begin{array}{rcl} \frac{1}{2} & = & \_\_ \\ +\ \frac{2}{3} & = & +\ \_\_ \\ \hline \end{array}$$

5. $$\begin{array}{rcl} \frac{4}{9} & = & \_\_ \\ -\ \frac{1}{6} & = & -\ \_\_ \\ \hline \end{array}$$

# Mixed Number Machine

Write the improper fraction for each mixed number. Use the corresponding letters to solve the riddle below.

1. $1\frac{2}{3}$ = ___ E
2. $2\frac{3}{4}$ = ___ G
3. $1\frac{1}{2}$ = ___ I
4. $2\frac{1}{8}$ = ___ M
5. $2\frac{1}{4}$ = ___ M
6. $3\frac{3}{8}$ = ___ A
7. $2\frac{2}{5}$ = ___ I
8. $1\frac{3}{5}$ = ___ P
9. $4\frac{3}{4}$ = ___ G

**"What do you call a pig that wants everything?"**

___    ___ ___ ___ ___ ___    ___ ___ ___

$\frac{27}{8}$    $\frac{11}{4}$ $\frac{12}{5}$ $\frac{17}{8}$ $\frac{9}{4}$ $\frac{5}{3}$    $\frac{8}{5}$ $\frac{3}{2}$ $\frac{19}{4}$

Change the improper fraction to a mixed number. Use the corresponding letters to solve the riddle below.

1. $\frac{7}{4}$ = A

2. $\frac{8}{3}$ = L

3. $\frac{3}{2}$ = P

4. $\frac{11}{8}$ = T

5. $\frac{12}{5}$ = N

6. $\frac{11}{6}$ = U

7. $\frac{21}{4}$ = E

8. $\frac{19}{16}$ = O

9. $\frac{19}{8}$ = C

10. $\frac{13}{2}$ = A

**"What kind of fruit can't run away and get married?"**

___ $2\frac{3}{8}$ ___ $1\frac{3}{4}$ ___ $2\frac{2}{5}$ ___ $1\frac{3}{8}$ ___ $6\frac{1}{2}$ ___ $2\frac{2}{3}$ ___ $1\frac{3}{16}$ ___ $1\frac{5}{6}$ ___ $1\frac{1}{2}$ ___ $5\frac{1}{4}$

## Take the Challenge

Find the sum. Write your answer as a mixed number in simplest terms. Use the corresponding letters to solve the riddle.

1. $3\frac{2}{4} + 2\frac{1}{4} =$ C
2. $5\frac{1}{6} + 5\frac{3}{6} =$ N
3. $5\frac{4}{12} + 4\frac{2}{12} =$ S
4. $8\frac{1}{8} + 2\frac{4}{8} =$ A
5. $6\frac{1}{3} + 3\frac{1}{3} =$ A
6. $4\frac{3}{16} + 7\frac{7}{16} =$ A
7. $7\frac{3}{8} + 4\frac{4}{8} =$ C
8. $6\frac{1}{8} + 2\frac{3}{8} =$ T

**"What kind of x-ray does a doctor use to examine a lion?"**

____ $9\frac{2}{3}$ ____ $11\frac{7}{8}$ ____ $10\frac{5}{8}$ ____ $8\frac{1}{2}$ ____ $9\frac{1}{2}$ ____ $5\frac{3}{4}$ ____ $11\frac{5}{8}$ ____ $10\frac{2}{3}$

Find the common denominator.
Regroup to make an improper fraction.
Subtract and write the difference as a mixed number in simplest terms.
Remember to show your work on a separate sheet of paper.

**Example:**

$$\begin{array}{rcl} 4\frac{1}{3} = 4\frac{2}{6} = 3\frac{8}{6} \\ -\ 1\frac{5}{6} = 1\frac{5}{6} = 1\frac{5}{6} \\ \hline 2\frac{3}{6} = 2\frac{1}{2} \end{array}$$

1. $$\begin{array}{r} 4\frac{2}{5} = \\ -\ 2\frac{7}{10} = \\ \hline \end{array}$$

2. $$\begin{array}{r} 7\frac{1}{5} = \\ -\ 3\frac{1}{2} = \\ \hline \end{array}$$

3. $$\begin{array}{r} 6\frac{1}{4} = \\ -\ 2\frac{1}{3} = \\ \hline \end{array}$$

4. $$\begin{array}{r} 5\frac{1}{4} = \\ -\ 2\frac{3}{4} = \\ \hline \end{array}$$

5. $$\begin{array}{r} 8\frac{1}{3} = \\ -\ 1\frac{5}{6} = \\ \hline \end{array}$$

# Diving Decimals

To multiply a fraction by a whole number, multiply the numerator by the whole number.

$$5 \times \frac{3}{7}$$

$$5 \times \frac{3}{7} = \frac{15}{7}$$

Then, if the product is an improper fraction, change it to a mixed number.

$$\frac{15}{7} = 2\frac{1}{7}$$

Multiply and show your work. Remember to simplify to the lowest terms.

1. $\frac{3}{4} \times 6 =$

2. $\frac{2}{3} \times 4 =$

3. $\frac{4}{5} \times 2 =$

4. $\frac{1}{2} \times 6 =$

5. $6 \times \frac{2}{5} =$

6. $9 \times \frac{1}{8} =$

7. $12 \times \frac{2}{3} =$

8. $20 \times \frac{1}{4} =$

9. $\frac{3}{5} \times 10 =$

10. $\frac{3}{8} \times 6 =$

Solve these problems. Show your work and don't forget to include the unit in your answer.

1. Mike is building a book case. He needs five boards that are each $2\frac{1}{4}$ feet long. Can Mike cut all five pieces out of one board that is 12 feet long? What is the total length of all five boards?

2. Mandy and Lindsey want to bake 8 loaves of bread for the baked-goods sale. The recipe for each loaf calls for $2\frac{1}{3}$ cups of flour. How much flour will they need?

3. Mrs. Brown is going to teach Mary and her friends how to sew a pillow. Each pillow takes $\frac{3}{4}$ of a yard of fabric. How many yards of fabric will she need to make 8 pillows?

# Mission Multiplication

List all the factors for each number.

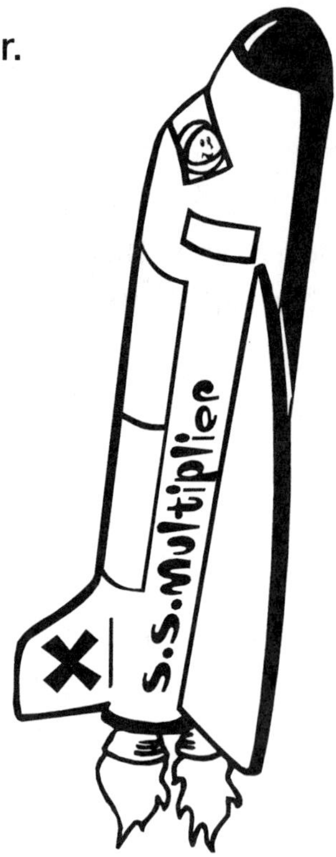


16 ______________________________

15 ______________________________

10 ______________________________

8 ______________________________

20 ______________________________

12 ______________________________

25 ______________________________

32 ______________________________

Name the greatest common factor for each pair.

16 and 32 ______

10 and 15 ______

8 and 20 ______

12 and 25 ______

# Answer Key

## Fun with Everyday Math

Grade 5

### Name That Number

Use < and > to compare each pair of numbers.

1. 3,642 < 4,693
2. 6,418 < 6,481
3. 21,009 > 20,999
4. 13,391 < 13,399
5. 2,791 < 2,971
6. 781,444 > 781,409
7. 1,109,322 > 1,108,999
8. 3,330,001 < 3,330,987

Circle the number that is . . .

9. less than 937. 937 939 (934)
10. greater than 4,671. 4,641 4,599 (4,700)
11. less than 21,413. (13,991) 21,415 21,600
12. less than 2,333 and greater than 2,105. 2,567 (2,222) 2,101



numeration comparing rounding

| State | Area in Square Miles |
|---|---|
| Illinois | 57,918 |
| Indiana | 36,420 |
| Iowa | 56,276 |
| Michigan | 96,705 |
| Minnesota | 86,943 |
| Ohio | 44,828 |

Round off the area of each of the states. Complete the chart.

| State | Round to the nearest . . . | | |
|---|---|---|---|
| | ten-thousand | thousand | hundred |
| Illinois | 60,000 | 58,000 | 57,900 |
| Indiana | 40,000 | 36,000 | 36,400 |
| Iowa | 60,000 | 56,000 | 56,300 |
| Michigan | 100,000 | 97,000 | 96,700 |
| Minnesota | 90,000 | 87,000 | 86,900 |
| Ohio | 40,000 | 45,000 | 44,800 |



### What's Missing?

Complete each of these number patterns.

1. 36 32 28 24 20 16 12 8 4 0
2. 21 28 35 42 49 56 63 70 77 84
3. 24 32 40 48 56 64 72 80 88 96
4. 36 45 54 63 72 81 90 99 108 117
5. 2 7 12 17 22 27 32 37 42 47
6. 41 46 51 56 61 66 71 76 81 86
7. 13 20 27 34 41 48 55 62 69 76
8. 120 118 116 114 112 110 108 106 104 102
9. 240 235 230 225 220 215 210 205
10. 103 106 109 112 115 118 121 124 127 130

number patterns

Study the numbers and complete the pattern.

99 118 137 156 175 194 213

121 117 113 109 105 101 97

400 406 412 418 424 430 436

842 836 830 824 818 812 806

224 221 218 215 212 209 206

207 214 221 228 235 242 249 256

1,025 1000 975 950 925 900 875 850

110 90 70 50 30 10 -10 -30

99 108 117 126 135 144 153 162

1,002 1,004 1,008 1,016 1032 1064 1128



### Just Write It!

| Simple number | | Expanded number |
|---|---|---|
| 41,673 | = | 40,000 + 1,000 + 600 + 70 + 3 |

Write each expanded number in its simple form.

1. 30,000 + 5,000 + 900 + 40 + 7 35,947
2. 60,000 + 800 + 50 + 6 60,856
3. 5,000 + 40 + 3 5,043
4. 80,000 + 3,000 + 80 + 6 83,086
5. 10,000 + 800 + 50 10,850

Write each simple number in its expanded form.

6. 32,463 30,000 + 2,000 + 400 + 60 + 3
7. 5,245 5,000 + 200 + 40 + 5
8. 3,006 3,000 + 6
9. 40,197 40,000 + 100 + 90 + 7
10. 9,100 9,000 + 100

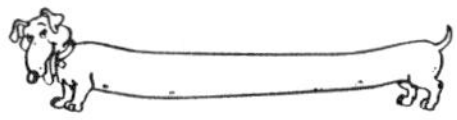



place value expanded word forms

| Numeral | Word Form |
|---|---|
| 26,874 = | twenty-six thousand, eight hundred seventy-four |

Write the numeral for each word form.

1. two thousand, five hundred sixty-one 2,561
2. twenty-three thousand, six hundred ninety-four 23,694
3. seventy-six thousand, eighty-six 76,086
4. four hundred thirteen thousand, one hundred seven 413,107
5. nine hunded forty-one thousand, one hundred ten 941,110

Write each simple number in word form.

6. 1,437 one thousand, four hundred thirty-seven
7. 21,900 twenty-one thousand, nine hundred
8. 80,063 eighty thousand, sixty-three
9. 10,013 ten thousand, thirteen
10. 123,004 one hundred twenty three thousand, four

## Estimation Game

Before you add, estimate the answer by rounding each number to its highest place value. After adding, compare the estimated sum with the actual sum. If your sum and estimation are not close, recheck your addition.

| Actual | Estimated |
|---|---|
| 392 | 400 |
| + 421 | + 400 |
| **813** | **800** |

1. 296 + 342 = 638; estimate 300 + 300 = 600
2. 4,726 + 3,121 = 7,847; estimate 5000 + 3000 = 8000
3. 782 + 413 = 1,195; estimate 800 + 400 = 1200
4. 2,827 + 8,293 = 11,120; estimate 3000 + 8000 = 11,000
5. 29 + 36 + 78 = 143; estimate 30 + 40 + 80 = 150
6. 85 + 97 + 23 = 205; estimate 90 + 100 + 20 = 210



Round each number to the nearest ten or hundred. Then multiply.

1. 296 x 18 = 6,000
2. 961 x 53 = 50,000
3. 423 x 72 = 28,000
4. 83 x 94 = 7,200
5. 623 x 595 = 360,000
6. 422 x 68 = 28,000
7. 32 x 38 = 1,200
8. 601 x 8 = 6,000

9. It is a 323 mile round trip between Grand Rapids, Michigan, and Chicago. The Amtrak train makes the round trip 62 times each month. Approximately how many miles does it travel in one month?

   300 x 60 = 18,000 miles

10. If the train averages 135 passengers on each round trip, and the train makes 62 round trips each month, approximately how many passengers make the trip each month?

    100 x 60 = 6,000 passengers



## Travel Yellowstone

The Williams family is taking a family vacation in Yellowstone National Park. They will drive to see many attractions in this large park.

Refer to the road map to calculate the shortest mileage they can travel between the locations listed.

First estimate the mileage between the two places named. Then calculate the actual mileage.

Estimates will vary.

| East Entrance and South Entrance | Madison Junction and Tower |
|---|---|
| Estimate — Actual: 70 mi. | Estimate — Actual: 44 mi. |
| **West Entrance and East Entrance** | **Northwest Entrance and South Entrance** |
| Estimate — Actual: 79 mi. | Estimate — Actual: 106 mi. |
| **West Entrance and Fishing Bridge** | **Old Faithful and Mammoth Hot Springs** |
| Estimate — Actual: 54 mi. | Estimate — Actual: 51 mi. |



## Subtraction Action

Rename the tens and show ten more ones. Then subtract.

**Example:** 463 − 246 = 217

Rename the hundreds and show ten more tens. Then subtract.

**Example:** 729 − 452 = 277

Find the difference.

1. 652 − 239 = 413
2. 847 − 493 = 354
3. 421 − 291 = 130
4. 4,683 − 2,591 = 2,092
5. 9,364 − 4,272 = 5,092
6. 9,601 − 7,281 = 2,320
7. 5,381 − 2,273 = 3,108
8. 7,736 − 5,728 = 2,008
9. 9,223 − 7,107 = 2,116



Regroup and subtract to find the difference.

**Examples:**

343 − 165 = 8 (partial) ; 343 − 165 = 178

1. 920 − 729 = 191
2. 856 − 298 = 558
3. 621 − 187 = 434
4. 867 − 268 = 599
5. 371 − 189 = 182
6. 863 − 378 = 485
7. 881 − 494 = 387
8. 641 − 258 = 383
9. 752 − 268 = 484



## Larry's Lawn and Garden Center

It's a warm, sunny Saturday in spring, and Larry's Lawn and Garden Center is a busy place. What will it cost each family to purchase its list of supplies for their lawns and gardens? Show your work. Remember you're computing money and label your answers appropriately. (There is no tax on the items.)

1. The Williams family needs 15 pounds of grass seed, two bags of fertilizer, and six rose bushes.
   $2.49 x 15 = $37.35; $6.25 x 2 = $12.50; $9.39 x 6 = $56.34; $37.35 + $12.50 + $56.34 = $106.19
2. The Johnson family wants to build a fence. They need 75 feet of wire fencing, eight fence posts, and six bags of cement.
   $.89 x 75 = $66.75; $5.76 x 8 = $46.08; $3.47 x 6 = $20.82; $66.75 + $46.08 + $20.82 = $133.65
3. The Ruiz family wants to plant a garden. They need eight tomato plants, eight tomato stakes, six flats of flowers, and two bags of fertilizer.
   $1.39 x 8 = $11.12; $.79 x 8 = $6.32; $6.79 x 6 = $40.74; $6.25 x 2 = $12.50; $11.12 + $6.32 + $40.74 + $12.50 = $70.68



## Measure Up!

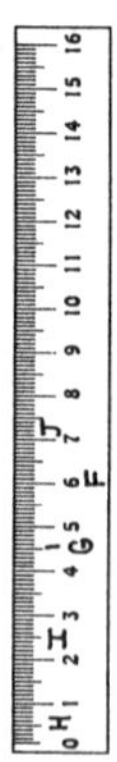

Mark each measurement on the inch ruler. The first one is done for you.

A. 1¾ inches
B. 5⅜ inches
C. 4⅞ inches
D. 3¼ inches
E. 2½ inches

Mark each measurement on the centimeter ruler.

F. 6 centimeters
G. 4½ centimeters
H. ½ centimeter
I. 25 millimeters
J. 72 millimeters

measurement U.S. customary metric

Circle the unit of measurement that you would use to measure each object.

Write the matching letters in the spaces at the bottom of the page to solve the riddle.

1. S kilometers / R meters / W centimeters (circled)

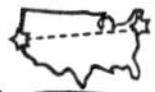

2. R kilometers (circled) / B meters / C centimeters

3. U meters / A millimeters (circled) / B centimeters

4. A miles / E feet (circled) / U inches

5. S inches / R miles / T yards (circled)

6. L inches (circled) / K feet / S yards

"What kind of water should you drink if you are sick?"

W E L L  W A T E R
1 4 6 6  1 3 5 4 2



## How Big Is It?

Determine the perimeter of each of these figures.

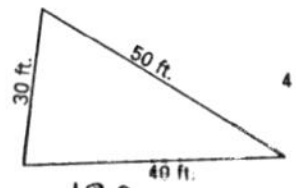


120 feet

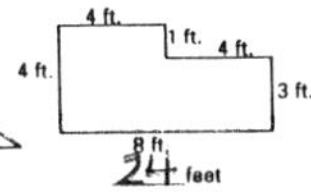


24 feet

1. Mr. Anderson's garden is in the shape of a 65' x 25' rectangle. How many feet of fencing would it take to enclose the garden?

   65 + 65 = 130; 25 + 25 = 50; 50 + 130 = 180 ft.

2. Coach Fritz told his team to run one time around the football field. The field is 360 feet long and 160 feet wide. How far do they have to run?

   360 + 360 = 720; 160 + 160 = 320; 320 + 720 = 1040 ft.



perimeter and area

Find the area of each figure in square feet.

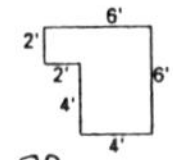


28 square feet

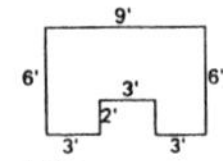


48 square feet

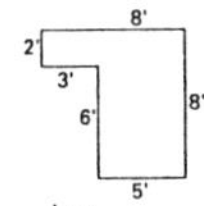


46 square feet

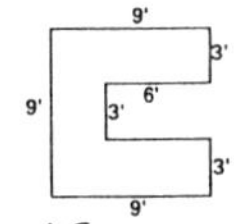


63 square feet



## Fraction Action

Write the fractional parts on each figure in simplest terms.

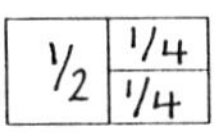


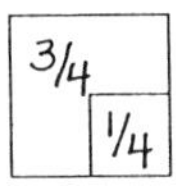


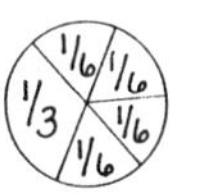


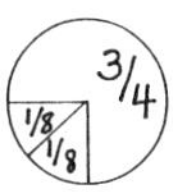


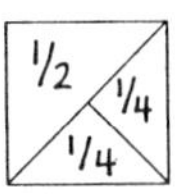


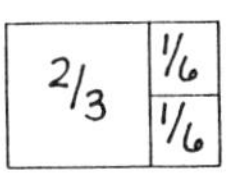




fractions naming simplifying

"If buttercups are yellow, what color are hiccups?"

BURPLE

Write each fraction in simplest form.

$\frac{3}{6} = \frac{1}{2}$  $\frac{9}{21} = \frac{3}{7}$  $\frac{6}{9} = \frac{2}{3}$

$\frac{5}{20} = \frac{1}{4}$  $\frac{16}{20} = \frac{4}{5}$  $\frac{10}{16} = \frac{5}{8}$

Cross out each simplified fraction and the letter above it. The remaining letters will solve the riddle.

| C (crossed out) | B | R (crossed out) | U | L (crossed out) | R | G (crossed out) | P | N (crossed out) | L | E | R (crossed out) |
|---|---|---|---|---|---|---|---|---|---|---|---|
| 1/2 | 3/6 | [illegible] | 6/10 | [illegible] | 4/12 | [illegible] | 6/9 | [illegible] | 2/4 | 4/8 | [illegible] |



## Another Name

Circle the fraction in each row that is not equivalent to the other four.

1. $\frac{1}{2}$ $\frac{2}{4}$ $\frac{6}{16}$ (circled) $\frac{3}{6}$ $\frac{16}{32}$
2. $\frac{5}{15}$ (circled) $\frac{2}{3}$ $\frac{8}{12}$ $\frac{6}{9}$ $\frac{16}{24}$
3. $\frac{9}{12}$ $\frac{6}{8}$ $\frac{3}{4}$ $\frac{15}{20}$ $\frac{20}{25}$ (circled)
4. $\frac{4}{16}$ $\frac{4}{12}$ (circled) $\frac{5}{20}$ $\frac{1}{4}$ $\frac{2}{8}$
5. $\frac{1}{5}$ $\frac{6}{60}$ (circled) $\frac{2}{10}$ $\frac{5}{25}$ $\frac{10}{50}$

Write three equivalent fractions for each given fraction.

1. $\frac{1}{3}$: $\frac{2}{6}$ $\frac{4}{12}$ $\frac{5}{15}$
2. $\frac{3}{5}$: $\frac{6}{10}$ $\frac{9}{15}$ $\frac{12}{20}$
3. $\frac{1}{10}$: $\frac{2}{20}$ $\frac{3}{30}$ $\frac{4}{40}$

Other answers possible.



equivalent fractions

Find the equivalent fraction.

1. $\frac{1}{2} = \frac{2}{4}$
2. $\frac{2}{3} = \frac{6}{9}$
3. $\frac{3}{4} = \frac{12}{16}$
4. $\frac{2}{8} = \frac{6}{24}$
5. $\frac{1}{5} = \frac{3}{15}$
6. $\frac{3}{8} = \frac{6}{16}$
7. $\frac{2}{10} = \frac{8}{40}$
8. $\frac{1}{4} = \frac{5}{20}$

## Add 'Em Up

Write the addition problem represented by each pair of fractions. Write the sum in simplest terms.

1. 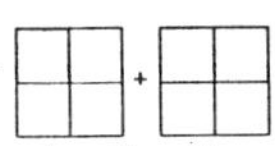

   $\frac{1}{4} + \frac{2}{4} = \frac{3}{4}$

2. 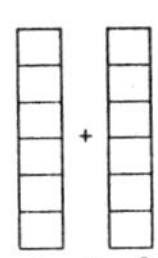

   $\frac{1}{6} + \frac{3}{6} = \frac{4}{6}$ or $\frac{2}{3}$

3. 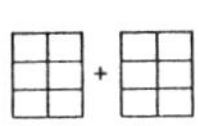

   $\frac{2}{6} + \frac{3}{6} = \frac{5}{6}$

4. 

   $\frac{1}{8} + \frac{4}{8} = \frac{5}{8}$

5. 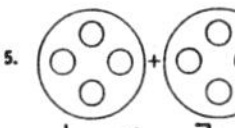

   $\frac{1}{4} + \frac{2}{4} = \frac{3}{4}$

6. 

   $\frac{1}{3} + \frac{2}{3} = \frac{3}{3}$ or 1



adding fractions

Find the sums. Write the sums in simplest terms and rename improper fractions as mixed numbers.

1. $\frac{7}{8} + \frac{4}{8} = \frac{11}{8} = 1\frac{3}{8}$
2. $\frac{3}{16} + \frac{7}{16} = \frac{10}{16} = \frac{5}{8}$
3. $\frac{1}{5} + \frac{3}{5} = \frac{4}{5}$
4. $\frac{4}{8} + \frac{5}{8} = \frac{9}{8} = 1\frac{1}{8}$
5. $\frac{5}{9} + \frac{2}{9} = \frac{7}{9}$
6. $\frac{3}{4} + \frac{3}{4} = \frac{6}{4} = 1\frac{1}{2}$
7. $\frac{2}{6} + \frac{1}{6} = \frac{3}{6} = \frac{1}{2}$
8. $\frac{6}{10} + \frac{8}{10} = \frac{14}{10} = 1\frac{2}{5}$
9. $\frac{3}{8} + \frac{1}{8} = \frac{4}{8} = \frac{1}{2}$
10. $\frac{7}{8} + \frac{3}{8} = \frac{10}{8} = 1\frac{1}{4}$

## Dancing Denominators

Solve by finding the least common denominator. Write the sum or difference in simplest terms.

**Example:** $\frac{3}{4} = \frac{6}{8}$, $-\frac{1}{8} = -\frac{1}{8}$, $\frac{5}{8}$

1. $\frac{1}{3} = \frac{2}{6}$, $+\frac{3}{6} = +\frac{3}{6}$, $\frac{5}{6}$
2. $\frac{7}{16} = \frac{7}{16}$, $-\frac{2}{8} = -\frac{4}{16}$, $\frac{11}{16}$
3. $\frac{1}{2} = \frac{4}{8}$, $-\frac{1}{8} = -\frac{1}{8}$, $\frac{5}{8}$
4. $\frac{2}{5} = \frac{4}{10}$, $+\frac{3}{10} = +\frac{3}{10}$, $\frac{7}{10}$
5. $\frac{1}{8} = \frac{2}{16}$, $+\frac{3}{16} = +\frac{3}{16}$, $\frac{5}{16}$



addition subtraction different denominators

Solve by finding the least common denominator. Write improper fractions as mixed numbers in their lowest terms.

**Example:** $\frac{4}{5} = \frac{12}{15}$, $-\frac{1}{3} = -\frac{5}{15}$, $\frac{7}{15}$

1. $\frac{3}{4} = \frac{9}{12}$, $+\frac{1}{3} = \frac{4}{12}$, $\frac{13}{12} = 1\frac{1}{12}$
2. $\frac{1}{3} = \frac{8}{24}$, $-\frac{2}{8} = -\frac{6}{24}$, $\frac{2}{24} = \frac{1}{12}$
3. $\frac{1}{6} = \frac{4}{24}$, $+\frac{3}{8} = +\frac{9}{24}$, $1\frac{3}{24}$
4. $\frac{1}{2} = \frac{3}{6}$, $+\frac{2}{3} = +\frac{4}{6}$, $\frac{7}{6} = 1\frac{1}{6}$
5. $\frac{4}{9} = \frac{8}{18}$, $-\frac{1}{6} = -\frac{3}{18}$, $\frac{5}{18}$

## Mixed Number Machine

Write the improper fraction for each mixed number. Use the corresponding letters to solve the riddle below.

1. $1\frac{2}{3} = \frac{5}{3}$ E
2. $2\frac{3}{4} = \frac{11}{4}$ G
3. $1\frac{1}{2} = \frac{3}{2}$ I
4. $2\frac{1}{8} = \frac{17}{8}$ M
5. $2\frac{1}{4} = \frac{9}{4}$ M
6. $3\frac{3}{8} = \frac{27}{8}$ A
7. $2\frac{2}{5} = \frac{12}{5}$ I
8. $1\frac{3}{5} = \frac{8}{5}$ P
9. $4\frac{3}{4} = \frac{19}{4}$ G

"What do you call a pig that wants everything?"

A GIMME-PIG

$\frac{27}{8}$   $\frac{11}{4}$ $\frac{12}{5}$ $\frac{17}{8}$ $\frac{9}{4}$ $\frac{5}{3}$   $\frac{8}{5}$ $\frac{3}{2}$ $\frac{19}{4}$



mixed numbers

Change the improper fraction to a mixed number. Use the corresponding letters to solve the riddle below.

1. $\frac{7}{4} = 1\frac{3}{4}$ A
2. $\frac{8}{3} = 2\frac{2}{3}$ L
3. $\frac{3}{2} = 1\frac{1}{2}$ P
4. $\frac{11}{8} = 1\frac{3}{8}$ T
5. $\frac{12}{5} = 2\frac{2}{5}$ N
6. $\frac{11}{6} = 1\frac{5}{6}$ U
7. $\frac{21}{4} = 5\frac{1}{4}$ E
8. $\frac{19}{16} = 1\frac{3}{16}$ O
9. $\frac{19}{8} = 2\frac{3}{8}$ C
10. $\frac{13}{2} = 6\frac{1}{2}$ A

"What kind of fruit can't run away and get married?"

CANTALOUPE

$2\frac{3}{8}$ $1\frac{3}{4}$ $2\frac{2}{5}$ $1\frac{3}{8}$ $6\frac{1}{2}$ $2\frac{2}{3}$ $1\frac{3}{16}$ $1\frac{5}{6}$ $1\frac{1}{2}$ $5\frac{1}{4}$



## Take the Challenge

Find the sum. Write your answer as a mixed number in simplest terms. Use the corresponding letters to solve the riddle.

1. $3\frac{2}{4} + 2\frac{1}{4} = 5\frac{3}{4}$ C
2. $5\frac{1}{6} + 5\frac{3}{6} = 10\frac{2}{3}$ N
3. $5\frac{4}{12} + 4\frac{2}{12} = 9\frac{1}{2}$ S
4. $8\frac{1}{8} + 2\frac{4}{8} = 10\frac{5}{8}$ A
5. $6\frac{1}{3} + 3\frac{1}{3} = 9\frac{2}{3}$ A
6. $4\frac{3}{16} + 7\frac{7}{16} = 11\frac{5}{8}$ A
7. $7\frac{3}{8} + 4\frac{4}{8} = 11\frac{7}{8}$ C
8. $6\frac{1}{8} + 2\frac{3}{8} = 8\frac{1}{2}$ T

"What kind of x-ray does a doctor use to examine a lion?"

A CAT-SCAN

$9\frac{2}{3}$   $11\frac{7}{8}$ $10\frac{5}{8}$ $8\frac{1}{2}$ $9\frac{1}{2}$ $5\frac{3}{4}$ $11\frac{5}{8}$ $10\frac{2}{3}$



addition subtraction mixed numbers

Find the common denominator. Regroup to make an improper fraction. Subtract and write the difference as a mixed number in simplest terms. Remember to show your work on a separate sheet of paper.

**Example:** $4\frac{1}{3} = 4\frac{2}{6} = 3\frac{8}{6}$, $-1\frac{5}{6} = 1\frac{5}{6} = 1\frac{5}{6}$, $2\frac{3}{6} = 2\frac{1}{2}$

1. $4\frac{2}{5} = 3\frac{14}{10}$, $-2\frac{7}{10} = 2\frac{7}{10}$, $1\frac{7}{10}$
2. $7\frac{1}{5} = 6\frac{12}{10}$, $-3\frac{1}{2} = 3\frac{5}{10}$, $3\frac{7}{10}$
3. $6\frac{1}{4} = 5\frac{15}{12}$, $-2\frac{1}{3} = 2\frac{4}{12}$, $3\frac{11}{12}$
4. $5\frac{1}{4} = 4\frac{10}{8}$, $-2\frac{3}{4} = 2\frac{6}{8}$, $2\frac{1}{2}$
5. $8\frac{1}{3} = 7\frac{8}{6}$, $-1\frac{5}{6} = 1\frac{5}{6}$, $6\frac{1}{2}$



## Diving Decimals

To multiply a fraction by a whole number, multiply the numerator by the whole number. $5 \times \frac{3}{7}$, $5 \times \frac{3}{7} = \frac{15}{7}$

Then, if the product is an improper fraction, change it to a mixed number. $\frac{15}{7} = 2\frac{1}{7}$

Multiply and show your work. Remember to simplify to the lowest terms.

1. $\frac{3}{4} \times 6 = \frac{18}{4} = 1\frac{1}{2}$
2. $\frac{2}{3} \times 4 = \frac{8}{3} = 2\frac{2}{3}$
3. $\frac{4}{5} \times 2 = \frac{8}{5} = 1\frac{3}{5}$
4. $\frac{1}{2} \times 6 = \frac{6}{2} = 3$
5. $6 \times \frac{2}{5} = \frac{12}{5} = 2\frac{2}{5}$
6. $9 \times \frac{1}{8} = \frac{9}{8} = 1\frac{1}{8}$
7. $12 \times \frac{2}{3} = \frac{24}{3} = 8$
8. $20 \times \frac{1}{4} = \frac{20}{4} = 5$
9. $\frac{3}{5} \times 10 = \frac{30}{5} = 6$
10. $\frac{3}{8} \times 6 = \frac{18}{8} = 2\frac{1}{4}$



fractions multiplying problem solving

Solve these problems. Show your work and don't forget to include the unit in your answer.

1. Mike is building a book case. He needs five boards that are each 2¼ feet long. Can Mike cut all five pieces out of one board that is 12 feet long? What is the total length of all five boards?
   Yes
   $5 \times 2\frac{1}{4} = \frac{45}{4} = 11\frac{1}{4}$ feet

2. Mandy and Lindsey want to bake 8 loaves of bread for the baked-goods sale. The recipe for each loaf calls for 2⅓ cups of flour. How much flour will they need?
   $8 \times 2\frac{1}{3} = \frac{7}{3} \times 8 = \frac{56}{3} = 18\frac{2}{3}$ cups

3. Mrs. Brown is going to teach Mary and her friends how to sew a pillow. Each pillow takes ¾ of a yard of fabric. How many yards of fabric will she need to make 8 pillows?
   $8 \times \frac{3}{4} = \frac{24}{4} = 6$ yards



## Mission Multiplication

List all the factors for each number.

- 16: 1, 2, 4, 8, 16
- 15: 1, 3, 5, 15
- 10: 1, 2, 5, 10
- 8: 1, 2, 4, 8
- 20: 1, 2, 4, 5, 10, 20
- 12: 1, 2, 3, 4, 6, 12
- 25: 1, 5, 25
- 32: 1, 2, 4, 8, 12, 16

Name the greatest common factor for each pair.

- 16 and 32: 16
- 10 and 15: 5
- 8 and 20: 4
- 12 and 25: 1

Complete the multiplication grids by multiplying each number along the top with each number along the left side. A few have been done for you.

| X | 6 | 3 | 8 | 4 |
|---|---|---|---|---|
| 2 | 12 | 6 | 16 | 8 |
| 7 | 42 | 21 | 56 | 28 |
| 3 | 18 | 9 | 24 | 12 |
| 9 | 54 | 27 | 72 | 36 |

| X | 7 | 4 | 9 | 8 |
|---|---|---|---|---|
| 7 | 49 | 28 | 63 | 56 |
| 9 | 63 | 36 | 81 | 72 |
| 3 | 21 | 12 | 27 | 24 |
| 5 | 35 | 20 | 45 | 40 |

| X | 9 | 8 | 6 | 7 |
|---|---|---|---|---|
| 5 | 45 | 40 | 30 | 35 |
| 8 | 72 | 64 | 48 | 56 |
| 2 | 18 | 16 | 12 | 14 |
| 4 | 36 | 32 | 24 | 28 |

## Fish for a Fact

Do you know your multiplication facts forwards, backwards, and inside out?

Find the missing number.

1. 6 x 6 = 36
2. 14 = 7 x 2
3. 4 x 7 = 28
4. 48 = 6 x 8
5. 6 x 9 = 54
6. 32 = 8 x 4
7. 8 x 9 = 72
8. 63 = 7 x 9
9. 5 x 9 = 45
10. 56 = 8 x 7
11. 7 x 7 = 49
12. 35 = 5 x 7
13. 8 x 6 = 48
14. 28 = 4 x 7
15. 7 x 9 = 63
16. 0 = 9 x 0
17. 4 x 4 = 16
18. 25 = 5 x 5
19. 3 x 9 = 27
20. 12 = 3 x 4

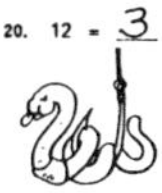



Try these two-step multiplication problems.

**Hint:** Look for combinations that make the second step easier. Sometimes it is easier to multiply the two numbers whose product equals a multiple of ten first.

**Examples:** 2 x 5 x 7 = ____ 2 x 5 = 10 then 10 x 7 = 70
3 x 3 x 9 = ____ 3 x 3 = 9 then 9 x 9 = 81

1. 2 x 2 x 7 = 28
2. 2 x 6 x 4 = 48
3. 7 x 6 x 2 = 84
4. 3 x 6 x 3 = 54
5. 4 x 4 x 4 = 64
6. 4 x 8 x 5 = 160
7. 9 x 2 x 5 = 90
8. 6 x 3 x 4 = 72
9. 7 x 2 x 3 = 42
10. 8 x 7 x 2 = 112
11. 8 x 4 x 5 = 160
12. 4 x 5 x 5 = 100
13. 6 x 3 x 2 = 36
14. 9 x 6 x 5 = 270
15. 5 x 7 x 8 = 280

## Product Practice

Find the product.

1. 20 x 3 = 60
2. 80 x 400 = 32,000
3. 400 x 30 = 12,000
4. 400 x 90 = 36,000
5. 20 x 50 = 1,000
6. 4 x 90 = 360
7. 60 x 6 = 360
8. 70 x 400 = 28,000
9. 70 x 30 = 2,100
10. 80 x 3 = 240
11. 700 x 40 = 28,000
12. 20 x 200 = 4,000
13. 600 x 30 = 18,000
14. 3 x 900 = 2,700
15. 40 x 70 = 2,800
16. 50 x 7 = 350
17. 60 x 70 = 4,200
18. 300 x 8 = 2,400
19. 80 x 6 = 480
20. 60 x 9 = 540

Find the product. Use the corresponding letters to solve the riddle below.

1. 57 x 3 = 171 A
2. 23 x 6 = 138 U
3. 47 x 5 = 235 S
4. 24 x 13 = 312 T
5. 32 x 96 = 3,072 I
6. 46 x 70 = 3,220 M
7. 214 x 431 = 92,234 D
8. 345 x 101 = 34,845 J
9. 723 x 80 = 57,840 K

**Q:** What did the little goat answer when the big goat asked, "What is 100 times 100?"

**A.** "How should I know?

I'M JUST A KID!"



## Double Digit Delight

Find the product. Then cross out the answers in the boxes. The remaining letters will answer the riddle.

1. 46 x 6 = 276
2. 87 x 3 = 261
3. 94 x 5 = 470
4. 84 x 6 = 504
5. 72 x 6 = 432
6. 98 x 3 = 294
7. 86 x 4 = 344
8. 57 x 6 = 342
9. 82 x 3 = 246
10. 91 x 7 = 637
11. 82 x 6 = 492
12. 21 x 7 = 147

| B | H | A | I | G | N |
|---|---|---|---|---|---|
| 637 | 701 | 492 | 633 | 294 | 504 |

| Y | A | S | N | U | A |
|---|---|---|---|---|---|
| 428 | 505 | 344 | 147 | 432 | 246 |

| B | G | U | S | P | D | R |
|---|---|---|---|---|---|---|
| 404 | 342 | 242 | 276 | 470 | 247 | 261 |

"What did Big Rose say to Little Rose?"

HI YA BUD



Find the products. Use the corresponding letters to solve the riddle below.

**Example:** 26 x 34 = 104 + 780 = 884

1. 37 x 42 = 1554 I
2. 63 x 71 = 4473 W
3. 48 x 63 = 3024 O
4. 80 x 55 = 4400 D
5. 53 x 42 = 2226 R
6. 49 x 23 = 1127 T
7. 13 x 39 = 507 G
8. 97 x 57 = 5529 L
9. 29 x 31 = 899 S
10. 42 x 74 = 3108 E
11. 82 x 16 = 1312 A
12. 78 x 98 = 7644 M

"What did you think of the Grand Canyon?"

IT WAS GORGES (1,554 1,127 4,473 1,312 899 507 3,024 2,226 507 3,108 899)



## Think It Through

Find the products. Use the corresponding letters to solve the riddle below.

**Example:** 264 x 412 = 528 + 2640 + 105600 = 108,768

1. 426 x 325 = 138,450 B
2. 601 x 312 = 187,512 L
3. 417 x 305 = 127,185 L
4. 512 x 438 = 224,256 U
5. 762 x 813 = 619,506 A
6. 209 x 914 = 191,026 T

"What's another name for a cowboy?"

A BULL (619,506 138,450 224,256 127,185 187,512)



Solve these story problems. Show your work and remember that the problems involve money.

1. Amanda babysits five days each week. She works for 3 hours each day. How much does she earn in one week if she charges $3.75 per hour? 5x3=15; $3.75 x 15 = $56.25
2. Matt, Eric, and Jamal did some landscaping work for Mr. Lubbers last week. Mr. Lubbers paid the boys a total of $217.50 for all their work. The boys split the money evenly. How much did each boy receive? $217.50 ÷ 3 = $72.50
3. Teisha painted 93 fire hydrants for the Fire Department last summer. She was paid $4.75 for each one that she painted. How much did she earn? $4.75 x 93 = $441.75
4. Tanya picked cherries at the Brown Fruit Farm last summer. She earned $0.45 for each quart that she picked. Tanya could pick 8 quarts in one hour. How much did she earn in one hour? 8 x $0.45 = $3.60

How much money could Tanya earn in an 8-hour day? 8 x $3.60 = $28.80

## Quotient Quota

Example: $3\overline{)24}$ = 8

Find the quotient.

1. $4\overline{)36}$ 9
2. $7\overline{)28}$ 4
3. $3\overline{)27}$ 9
4. $4\overline{)32}$ 8
5. $6\overline{)48}$ 8
6. $7\overline{)56}$ 8
7. $9\overline{)72}$ 8
8. $9\overline{)81}$ 9
9. $5\overline{)45}$ 9
10. $4\overline{)12}$ 3
11. $6\overline{)24}$ 4
12. $7\overline{)49}$ 7
13. $3\overline{)12}$ 4
14. $5\overline{)30}$ 6
15. $7\overline{)21}$ 3
16. $3\overline{)18}$ 6
17. $8\overline{)64}$ 8
18. $7\overline{)42}$ 6
19. $8\overline{)40}$ 5
20. $6\overline{)36}$ 6
21. $6\overline{)24}$ 4
22. $3\overline{)21}$ 7
23. $4\overline{)28}$ 7
24. $8\overline{)72}$ 9



division with one digit divisor

Find the quotient. Cross out the quotients in the boxes. The remaining letters will spell the answer to the riddle.

1. $5\overline{)95}$ 19
2. $6\overline{)852}$ 142
3. $4\overline{)940}$ 235
4. $7\overline{)938}$ 134
5. $3\overline{)762}$ 254
6. $4\overline{)864}$ 216
7. $3\overline{)744}$ 248
8. $8\overline{)976}$ 122
9. $5\overline{)965}$ 193

| 485 A | 122 I | 19 T | | | | | | | | | |
|---|---|---|---|---|---|---|---|---|---|---|---|
| 216 S | 532 G | 121 I | 142 U | 322 R | 345 R | 235 P | 14 A | 18 F | 248 A | 238 I | 242 C |
| 193 B | 195 J | 134 E | 342 A | 254 T | 35 M | | | | | | |

"What do you call it when giraffes going in one direction get mixed-up with giraffes going in another direction?"

A GIRRAFIC JAM

## Extend Yourself with Division

Complete these division extensions.

15 ÷ 3 = 5
150 ÷ 3 = 50
1,500 ÷ 3 = 500
15,000 ÷ 3 = 5,000

28 ÷ 7 = 4
2,800 ÷ 7 = 400
28,000 ÷ 7 = 4,000

36 ÷ 6 = 6
360 ÷ 6 = 60
36,000 ÷ 6 = 6,000

480 ÷ 8 = 60
4,800 ÷ 8 = 600
48,000 ÷ 8 = 6,000

Write your own extensions to these division problems.

18 ÷ 6 = 3 | 24 ÷ 6 = 4
180 ÷ 6 = 30 | 240 ÷ 6 = 40
1800 ÷ 6 = 300 | 2,400 ÷ 6 = 400
18,000 ÷ 6 = 3,000 | 24,000 ÷ 6 = 4,000



division extensions

Find the quotients for these division extensions.

54 ÷ 9 = 6
540 ÷ 90 = 6
5,400 ÷ 900 = 6
54,000 ÷ 9,000 = 6

42 ÷ 6 = 7
420 ÷ 60 = 7
4,200 ÷ 600 = 7
42,000 ÷ 6,000 = 7

32 ÷ 8 = 4
320 ÷ 80 = 4
3,200 ÷ 800 = 4

27 ÷ 9 = 3
270 ÷ 9 = 3
2,700 ÷ 90 = 3
27,000 ÷ 900 = 3

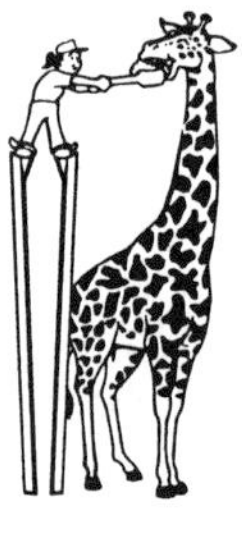

Make two sets of division extension problems of your own.

ANSWERS WILL VARY.



## Divide and Conquer

Example: quotient → 15 R 3 ← remainder; divisor → $4\overline{)63}$ ← dividend; 4; 23; 20; 3

Find the quotient.

1. $4\overline{)175}$ 43 r.3
2. $7\overline{)253}$ 36 r.1
3. $5\overline{)63}$ 12 r.3
4. $3\overline{)89}$ 29 r.2
5. $7\overline{)99}$ 14 r.1
6. $8\overline{)93}$ 11 r.5
7. $6\overline{)153}$ 25 r.3
8. $9\overline{)128}$ 14 r.2
9. $8\overline{)146}$ 18 r.2



division with remainders

To check your work, multiply the quotient by the divisor. Then add the remainder.

Example: $16\overline{)36}$ = 2 R 4; 32; 4 — 16 × 2 = 32; + 4 = 36

1. $23\overline{)96}$ 4 r.4 — 23 × 4 = 92; + 4 = 96
2. $16\overline{)86}$ 5 r.6 — 16 × 5 = 80; + 6 = 86
3. $27\overline{)93}$ 3 r.12 — 27 × 3 = 81; + 12 = 93
4. $13\overline{)88}$ 6 r.10 — 13 × 6 = 78; + 10 = 88
5. $24\overline{)753}$ 31 r.9 — 24 × 31 = 24 + 72 = 744; + 9 = 753
6. $48\overline{)687}$ 14 r.15 — 48 × 14 = 192 + 48 = 672; + 15 = 687



## Determining Decimals

Match the decimals. Write the letters in the blanks below to answer the riddle. The first one is done for you.

1. three tenths — 5.62 U
2. two and four tenths — 0.03 M
3. five and sixty-two hundredths — 0.003 G
4. three hundredths — 0.3 A
5. fifty-six and two tenths — 5.062 N
6. two and four thousandths — 0.562 R
7. five and sixty-two thousandths — 2.4 H
8. three thousandths — 2.040 E
9. two and forty thousandths — 2.004 I
10. five hundred sixty-two thousandths — 56.2 D

"What do you get when you cross a hummingbird with a doorbell?"

A HUMDINGER (1–10)



reading writing decimals

Write these words in decimal form.

1. two and four tenths 2.4
2. sixteen and twenty-seven hundredths 16.27
3. eight and four hundredths 8.04
4. twenty-six and seven tenths 26.7
5. nineteen hundredths 0.19

Write these decimals in words.

6. 5.3 five and three tenths
7. 12.15 twelve + fifteen hundredths
8. 1.06 one + six hundredths
9. 2.3 two + three tenths
10. 3.03 three + three hundredths
11. 100.0 one hundred
12. 12.77 twelve + seventy seven hundredths

## How Do They Compare?

Circle the largest number in each row.

1. G 0.031 | C 0.13 | (A 0.8)
2. B 1.01 | (H 1.10) | L 1.010
3. (A 2.301) | E 2.056 | U 2.30
4. B 0.709 | (T 0.76) | R 0.09

Circle the smallest number in each row.

5. (C 0.001) | K 0.01 | T 0.100
6. (H 5.03) | P 5.1 | R 5.031
7. E 9.10 | U 9.70 | (I 9.07)
8. R 7.001 | (T 7.0) | B 7.20

"What kind of tool does a baby bird need for getting out of an eggshell?"

Write the circled letter from each problem above.

A HATCH-IT (1–8)

Write <, >, or = to compare the decimals below. Circle the letter under the symbol used in each problem.

| | | | < | > | = |
|---|---|---|---|---|---|
| 1. 4.248 | < | 4.261 | (C) | T | R |
| 2. 41.01 | = | 41.010 | O | U | (R) |
| 3. 6.035 | > | 6.008 | K | (O) | P |
| 4. 5.045 | < | 5.45 | (A) | K | B |
| 5. 4.26 | < | 4.62 | (K) | O | W |
| 7. 0.07 | < | 0.7 | (A) | M | C |
| 6. 0.10 | = | 0.100 | I | E | (C) |
| 8. 4.301 | < | 4.310 | (O) | V | S |
| 9. 5.030 | = | 5.03 | L | R | (L) |
| 10. 0.9 | > | 0.099 | A | (A) | L |

Write the circled letters from the chart above in order in the spaces below to answer the riddle.

"What do frogs drink at picnics?"

C R O A K A C O L A
1 2 3 4 5 6 7 8 9 10



## Order of the Day

List the numbers from least to greatest.

1. 3.62 3.32 3.091 — 3.091 3.32 3.62
2. 5.3 5.6 5.5 — 5.3 5.5 5.6
3. 71.20 71.38 71.3 — 71.20 71.3 71.38
4. 3.01 3.10 3.011 — 3.01 3.011 3.10
5. 1.7 1.723 1.72 — 1.7 1.72 1.723

Fill in the missing numbers.

6. 0.07 0.08 0.09 0.10 0.11 0.12 0.13 0.14
7. 3.4 3.5 3.6 3.7 3.8 3.9 4.0
8. 8.27 8.28 8.29 8.30 8.31 8.32 8.33 8.34
9. 3.9 4.0 4.1 4.2 4.3 4.4 4.5
10. 0.007 0.008 0.009 0.010 0.011 0.012

Order the numbers from least to greatest on the lines below. Then write the corresponding letters in the boxes.

| | | | | | |
|---|---|---|---|---|---|
| 3.6 | I | 3.00 | I | 2.1 | A |
| 4.01 | O | 3.61 | D | 4.45 | N |
| 2.111 | C | 4.09 | A | 2.9 | H |
| 4.06 | N | 3.999 | G | 3.9 | O |
| 3.4 | L | 4.2 | B | 4.4 | U |

"What do you call a cold puppy that is sitting on a baby rabbit?"

A C H I L I — 2.1 2.111 2.9 3.00 3.4 3.6

D O G O N — 3.61 3.9 3.999 4.01 4.06

A B U N — 4.09 4.2 4.4 4.45



## Round the Corner

Complete the chart by rounding each number as directed.

| | Whole Number | Tenth | Hundredth |
|---|---|---|---|
| 6.013 | 6 | 6.0 | 6.01 |
| 23.278 | 23 | 23.3 | 23.28 |
| 4.496 | 4 | 4.5 | 4.50 |
| 8.021 | 8 | 8.0 | 8.02 |
| 9.836 | 10 | 9.8 | 9.84 |
| 4.997 | 5 | 5.0 | 5.00 |
| 5.877 | 6 | 5.9 | 5.88 |
| 7.391 | 7 | 7.4 | 7.39 |
| 10.099 | 10 | 10.1 | 10.10 |
| 100.990 | 101 | 101.0 | 100.99 |

Round each decimal as directed. Cross out the boxes with the correct answers. The remaining letters will answer the riddle below.

Round . . .

1. 4.09 to the nearest tenth. 4.1
2. 24.31 to the nearest whole number. 24
3. 3.066 to the nearest hundredth. 3.07
4. 4.33 to the nearest tenth. 4.3
5. 5.88 to the nearest tenth. 5.9
6. 4.893 to the nearest hundredth. 4.89

| | | | | | | | |
|---|---|---|---|---|---|---|---|
| M 3.08 | ~~B 4.1~~ | I 4.90 | L 4.4 | ~~C 24~~ | K 4.0 | ~~A 4.89~~ | S 4.0 |
| ~~B 5.9~~ | N 25 | A 3.1 | ~~G 3.07~~ | K 4.09 | E 5.8 | ~~R 4.3~~ | S 5.6 |

What kind of reptiles do kids like to drink with their hamburgers?

M I L K
S N A K E S

## Sum Decimals

Find the sums. Write the corresponding letters in the riddle below.

1. 0.35 + 0.7 = 1.05 L
2. 3.816 + 2.719 = 6.535 U
3. 7.1 + 32.90 = 40.00 O
4. 6.831 + 4.9 = 11.731 I
5. 5.7 + 2.368 = 8.068 T
6. 2.010 + 99.99 = 102.00 M
7. 0.213 + 0.937 = 1.150 S
8. 3.446 + 1.71 = 5.156 P
9. 1.8 + 0.09 = 3.69 F

"What kind of flowers grow under your nose?"

T U L I P S
8.068 6.535 1.05 11.731 5.156 1.150



Find the sum. Cross out the answers in the boxes. The remaining letters will answer the ridddle.

1. 3.41 + 2.601 = 6.011
2. 4.63 + 1.033 = 5.663
3. 3.8 + 2.012 = 5.812
4. 2.61 + 2.7 = 5.31
5. 3.6 + 0.47 = 4.07
6. 0.011 + 3.67 = 3.681

| | | | | | | |
|---|---|---|---|---|---|---|
| M 3.40 | ~~B 3.681~~ | I 3.13 | N 5.41 | ~~R 6.011~~ | U 6.1 | ~~C 5.663~~ |
| T 7.2 | E 8.9 | ~~G 5.31~~ | M 3.99 | I 1.71 | ~~K 4.07~~ | C 2.76 |
| ~~Y 5.812~~ | E 3.00 | | | | | |

"What do cats eat when they are in a hurry?"

M I N U T E
M I C E



## Take It Away

Find the differences.

1. 61.32 − 42.3 = 19.02
2. 4.813 − 2.32 = 2.493
3. 5.8 − 4.13 = 1.67
4. 14.24 − 1.031 = 13.209
5. 54.27 − 3.36 = 50.91
6. 0.581 − 0.19 = 0.391
7. 14.3 − 3.21 = 11.09
8. 6.53 − 1.3 = 5.23

Find the sum or difference. Show your work for the number stories.

1. $31.47 + 16.21 + 18.37 = $66.05
2. $47.03 − 12.85 = $34.18
3. $1.35 + $12.23 = $13.58
4. $16.27 + $8.15 = $8.12
5. Last summer Chris earned $83.75 mowing lawns, $23.50 washing cars, and $68.45 babysitting. How much did he earn from all three jobs?

$83.75 + $23.50 + 68.45 = $175.70

6. How much more did Chris earn from mowing lawns than from babysitting?

$83.75 − 68.45 = $15.30

## Land and Sea

Find the products. Use the corresponding letters to solve the riddle below.

1. 3.61 × 5 = 18.05 H
2. 4.27 × 8 = 34.16 R
3. 6.42 × 3 = 19.26 I
4. 2.31 × 21 = 48.51 C
5. 3.70 × 43 = 159.10 I
6. 0.081 × 36 = 2.916 L
7. 6.17 × 314 = 1937.38 S
8. 8.21 × 476 = 3907.96 O
9. 6.322 × 57 = 360.354 G

"What do you get when you bury money in your garden?"

| R | I | C | H |
|---|---|---|---|
| 34.16 | 159.10 | 48.51 | 18.05 |

| S | O | I | L |
|---|---|---|---|
| 1,937.38 | 3,907.96 | 19.26 | 2.916 |



---

Find the quotients. Show your work. Then use the corresponding letters to solve the riddle below.

1. 25.92 ÷ 6 = 4.32 S
2. 94.8 ÷ 4 = 23.7 W
3. 1.875 ÷ 5 = 0.375 A
4. 37.38 ÷ 7 = 5.34 E
5. 1.449 ÷ 3 = 0.483 S
6. 280.8 ÷ 8 = 35.1 A

"What do you use to divide the ocean in half?"

A

| S | E | A | - | S | A | W |
|---|---|---|---|---|---|---|
| 0.483 | 5.34 | 0.375 | | 4.32 | 35.1 | 23.7 |



---

## Zero Zonkers

Find the products. Use the corresponding letters to solve the riddle below.

1. 13.4 × 0.2 = 2.68 M
2. 8.7 × 5 = 43.5 N
3. 1.63 × 0.4 = 0.652 E
4. 3.4 × 6 = 20.4 A
5. 214 × 0.3 = 64.2 L
6. 9.9 × 4 = 39.6 O
7. 2.3 × 0.4 = 0.92 I
8. 3.31 × 4 = 13.24 S
9. 10.1 × 0.5 = 5.05 G
10. 4.6 × 3 = 13.8 T

"What did the little crab say when the tide finally came in?"

| L | O | N | G | T | I | M | E |
|---|---|---|---|---|---|---|---|
| 64.2 | 39.6 | 43.5 | 5.05 | 13.8 | 0.92 | 2.68 | 0.652 |

| N | O | S | E | A |
|---|---|---|---|---|
| 43.5 | 39.6 | 13.24 | 0.652 | 20.4 |



---

Find the product. Use the corresponding letters to solve the riddle below.

1. 0.76 × 1.3 = 0.988 U
2. 6.13 × 0.05 = 0.3065 L
3. 24.6 × 2.3 = 56.58 E
4. 68.1 × 0.04 = 2.724 A
5. 34.3 × 0.14 = 4.802 P
6. 7.8 × 0.03 = 0.234 D

"What do you call a snowman on a warm day?"

| A | P | U | D | D | L | E |
|---|---|---|---|---|---|---|
| 2.724 | 4.802 | 0.988 | 0.234 | 0.234 | 0.3065 | 56.58 |



---

## Take Me Out to the Ballgame

| Souvenir Stand | |
|---|---|
| Autographed Baseballs | $24.95 |
| Baseball Hats | $12.75 |
| Programs | $2.50 |
| Pennants | $4.75 |
| T-shirts | $13.49 |
| Sweatshirts | $22.25 |
| Batting Helmets | $4.50 |
| Water Bottle | $2.75 |
| Mini Poster | $3.50 |
| Life Size Poster | $9.50 |
| Coloring Book | $2.25 |

---

Use the chart on page 68 to solve these money problems. Remember to show your work and label the answers.

1. Kyle wants to buy a mini-poster for each of his three little sisters. How much will it cost for 3 posters? Kyle has a $20 bill. How much change will Kyle receive?

   $3.50 × 3 = $10.50; $20.00 − $10.50 = $9.50

2. Mr. Matthews has $50 to spend on prizes for the school candy sale. He wants to buy an autographed baseball, a helmet, a T-shirt, and a pennant. How much will he spend?

   $24.95 + $4.50 = $29.45; $13.49 + $4.75 = $18.24; $18.24 + $29.45 = $47.69

   With the change from his $50, what else can Mr. Matthews buy from the souvenir stand?

   A coloring book

3. The souvenir stand sold 42 sweatshirts on a cold day at the ballpark. How much money did they take in from the sale of the sweatshirts?

   $22.25 × 42 = $934.50

   The souvenir stand bought the sweatshirts for $12 each. How much profit did they make on each sweatshirt? How much profit did they make on the 42 sweatshirts?

   $22.25 − $12. = $10.25



---

## Chart It

The Eagles played 6 basketball games. Find the total points scored per game and the season average for each player.

| Game in Series | | | | | | | |
|---|---|---|---|---|---|---|---|
| Player | 1 | 2 | 3 | 4 | 5 | 6 | Player Averages |
| Laura | 2 | 3 | 4 | 2 | 3 | 4 | 3 |
| Mandy | 5 | 4 | 4 | 6 | 8 | 9 | 6 |
| Jan | 12 | 10 | 13 | 6 | 7 | 12 | 10 |
| Nancy | 6 | 12 | 8 | 5 | 3 | 8 | 7 |
| Mary | 2 | 2 | 10 | 12 | 10 | 0 | 6 |
| Amy | 10 | 4 | 3 | 8 | 4 | 7 | 6 |
| Carol | 5 | 2 | 1 | 3 | 0 | 7 | 3 |
| Game Totals | 42 | 37 | 43 | 42 | 35 | 47 | |

---

Use each rule to solve the problems and complete the charts.

| N | Rule: N + 5 |
|---|---|
| 3 | 3 + 5 = 8 |
| 6 | 6+5=11 |
| 9 | 9+5=14 |
| 12 | 12+5=17 |
| 16 | 16+5=21 |

| N | Rule: N x 4 |
|---|---|
| 3 | 3x4=12 |
| 5 | 5x4=12 |
| 7 | 7x4=28 |
| 8 | 8x4=32 |
| 9 | 9x4=36 |

| N | Rule: (n + 2) x 3 |
|---|---|
| 6 | (6+2)x3=24 |
| 8 | (8+2)x3=30 |
| 10 | (10+2)x3=36 |
| 5 | (5+2)x3=21 |
| 4 | (4+2)x3=18 |

| N | Rule: (n + 3) ÷ 4 |
|---|---|
| 17 | (17+3)÷4=5 |
| 1 | (1+3)÷4=1 |
| 13 | (13+3)÷4=4 |
| 33 | (33+3)÷4=9 |
| 5 | (5+3)÷4=2 |



---

## Mystery Number

Here is a little trick that you can do with your house number and your age.

First try it with your own house number and age. Then try it out on a friend.

Example:

| | |
|---|---|
| Write your house number. | 2065 |
| Multiply by 2. | 4130 |
| Add 5. | 4135 |
| Multiply by 50. | 206750 |
| Add your age in years. (12) | 206762 |
| Add 365. | 207127 |
| Subtract 615. | 206512 |

Can you find your house number and age in the answer?

Answers will vary.

Complete the multiplication grids by multiplying each number along the top with each number along the left side. A few have been done for you.

| X | 6 | 3 | 8 | 4 |
|---|---|---|---|---|
| 2 | 12 | | | |
| 7 | | | 56 | |
| 3 | 18 | | | |
| 9 | | | | |

| X | 7 | 4 | 9 | 8 |
|---|---|---|---|---|
| 7 | | | | |
| 9 | | | | |
| 3 | | | | |
| 5 | | | | |

| X | 9 | 8 | 6 | 7 |
|---|---|---|---|---|
| 5 | | | | |
| 8 | | | | |
| 2 | | | | |
| 4 | | | | |

# Fish for a Fact

Do you know your multiplication facts forwards, backwards, and inside out?

Find the missing number.

1. ____ x 6 = 36
2. 14 = ____ x 2
3. 4 x ____ = 28
4. ____ = 6 x 8
5. 6 x 9 = ____
6. 32 = ____ x 4
7. 8 x ____ = 72
8. 63 = ____ x 9
9. 5 x 9 = ____
10. 56 = 8 x ____
11. ____ x 7 = 49
12. ____ = 5 x 7
13. ____ x 6 = 48
14. 28 = ____ x 7
15. 7 x 9 = ____
16. 0 = 9 x ____
17. 4 x ____ = 16
18. 25 = 5 x ____
19. 3 x ____ = 27
20. 12 = ____ x 4

Try these two-step multiplication problems.

**Hint:** Look for combinations that make the second step easier. Sometimes it is easier to multiply the two numbers whose product equals a multiple of ten first.

**Examples:** 2 x 5 x 7 = _____ 2 x 5 = 10 then 10 x 7 = 70

3 x 3 x 9 = _____ 3 x 3 = 9 then 9 x 9 = 81

1. 2 x 2 x 7 = _____
2. 2 x 6 x 4 = _____
3. 7 x 6 x 2 = _____
4. 3 x 6 x 3 = _____
5. 4 x 4 x 4 = _____
6. 4 x 8 x 5 = _____
7. 9 x 2 x 5 = _____
8. 6 x 3 x 4 = _____
9. 7 x 2 x 3 = _____
10. 8 x 7 x 2 = _____
11. 8 x 4 x 5 = _____
12. 4 x 5 x 5 = _____
13. 6 x 3 x 2 = _____
14. 9 x 6 x 5 = _____
15. 5 x 7 x 8 = _____

# Product Practice

Find the product.

1. 20 x 3 = ____________ 
2. 80 x 400 = ____________
3. 400 x 30 = ____________
4. 400 x 90 = ____________
5. 20 x 50 = ____________
6. 4 x 90 = ____________
7. 60 x 6 = ____________
8. 70 x 400 = ____________
9. 70 x 30 = ____________
10. 80 x 3 = ____________
11. 700 x 40 = ____________
12. 20 x 200 = ____________
13. 600 x 30 = ____________
14. 3 x 900 = ____________
15. 40 x 70 = ____________
16. 50 x 7 = ____________
17. 60 x 70 = ____________
18. 300 x 8 = ____________
19. 80 x 6 = ____________
20. 60 x 9 = ____________

Find the product. Use the corresponding letters to solve the riddle below.

| | | |
|---|---|---|
| 1. $57 \times 3$ = ____ A | 2. $23 \times 6$ = ____ U | 3. $47 \times 5$ = ____ S |
| 4. $24 \times 13$ = ____ T | 5. $32 \times 96$ = ____ I | 6. $46 \times 70$ = ____ M |
| 7. $214 \times 431$ = ____ D | 8. $345 \times 101$ = ____ J | 9. $723 \times 80$ = ____ K |

**Q:** What did the little goat answer when the big goat asked, "What is 100 times 100?"

**A.** "How should I know?

____ ____' ____ ____ ____ ____
3,072 3,220 34,845 138 235 312

____ ____ ____ ____!"
171 57,840 3,072 92,234

# Double Digit Delight

Find the product. Then cross out the answers in the boxes. The remaining letters will answer the riddle.

| | | | |
|---|---|---|---|
| 1. 46 x 6 | 2. 87 x 3 | 3. 94 x 5 | 4. 84 x 6 |
| 5. 72 x 6 | 6. 98 x 3 | 7. 86 x 4 | 8. 57 x 6 |
| 9. 82 x 3 | 10. 91 x 7 | 11. 82 x 6 | 12. 21 x 7 |

| B | H | A | I | G | N |
|---|---|---|---|---|---|
| 637 | 701 | 492 | 633 | 294 | 504 |

| Y | A | S | R | U | A |
|---|---|---|---|---|---|
| 428 | 505 | 344 | 147 | 432 | 246 |

| B | G | U | S | P | D | R |
|---|---|---|---|---|---|---|
| 404 | 342 | 242 | 276 | 470 | 247 | 261 |

**"What did Big Rose say to Little Rose?"**

____ ____   ____ ____   ____ ____ ____

Find the products. Use the corresponding letters to solve the riddle below.

**Example:**

$$\begin{array}{r} {}^{1}_{2}\\ 26 \\ \times\ 34 \\ \hline 104 \\ +780 \\ \hline 884 \end{array}$$

1. $\begin{array}{r} 37 \\ \times 42 \\ \hline \end{array}$ I

2. $\begin{array}{r} 63 \\ \times 71 \\ \hline \end{array}$ W

3. $\begin{array}{r} 48 \\ \times 63 \\ \hline \end{array}$ O

4. $\begin{array}{r} 80 \\ \times 55 \\ \hline \end{array}$ D

5. $\begin{array}{r} 53 \\ \times 42 \\ \hline \end{array}$ R

6. $\begin{array}{r} 49 \\ \times 23 \\ \hline \end{array}$ T

7. $\begin{array}{r} 13 \\ \times 39 \\ \hline \end{array}$ G

8. $\begin{array}{r} 97 \\ \times 57 \\ \hline \end{array}$ L

9. $\begin{array}{r} 29 \\ \times 31 \\ \hline \end{array}$ S

10. $\begin{array}{r} 42 \\ \times 74 \\ \hline \end{array}$ E

11. $\begin{array}{r} 82 \\ \times 16 \\ \hline \end{array}$ A

12. $\begin{array}{r} 78 \\ \times 98 \\ \hline \end{array}$ M

**"What did you think of the Grand Canyon?"**

____ ____    ____ ____ ____
1,554 1,127    4,473 1,312 899

____ ____ ____ ____ ____ ____
507 3,024 2,226 507 3,108 899

1 − 2 3 + − 3 2 = 1 + 2 3

# Think It Through

Find the products. Use the corresponding letters to solve the riddle below.

**Example:**

```
    2 1
    1
    264
  x 412
    528
   2640
 105600
108,768
```

1.
```
  426
x 325
```
B

2.
```
  601
x 312
```
L

3.
```
  417
x 305
```
L

4.
```
  512
x 438
```
U

5.
```
  762
x 813
```
A

6.
```
  209
x 914
```
T

**"What's another name for a cowboy?"**

_____ 619,506

_____ 138,450 _____ 224,256 _____ 127,185 _____ 187,512

Solve these story problems. Show your work and remember that the problems involve money.

1. Amanda babysits five days each week. She works for 3 hours each day. How much does she earn in one week if she charges $3.75 per hour?

2. Matt, Eric, and Jamal did some landscaping work for Mr. Lubbers last week. Mr. Lubbers paid the boys a total of $217.50 for all their work. The boys split the money evenly. How much did each boy receive?

3. Teisha painted 93 fire hydrants for the Fire Department last summer. She was paid $4.75 for each one that she painted. How much did she earn?

4. Tanya picked cherries at the Brown Fruit Farm last summer. She earned $0.45 for each quart that she picked. Tanya could pick 8 quarts in one hour. How much did she earn in one hour?

   How much money could Tanya earn in an 8-hour day?

# Quotient Quota

**Example:** $\begin{array}{r} 8 \\ 3\overline{)24} \end{array}$

Find the quotient.

1. $4\overline{)36}$ 2. $7\overline{)28}$ 3. $3\overline{)27}$ 4. $4\overline{)32}$

5. $6\overline{)48}$ 6. $7\overline{)56}$ 7. $9\overline{)72}$ 8. $9\overline{)81}$

9. $5\overline{)45}$ 10. $4\overline{)12}$ 11. $6\overline{)24}$ 12. $7\overline{)49}$

13. $3\overline{)12}$ 14. $5\overline{)30}$ 15. $7\overline{)21}$ 16. $3\overline{)18}$

17. $8\overline{)64}$ 18. $7\overline{)42}$ 19. $8\overline{)40}$ 20. $6\overline{)36}$

21. $6\overline{)24}$ 22. $3\overline{)21}$ 23. $4\overline{)28}$ 24. $8\overline{)72}$

division with one digit divisor

Find the quotient. Cross out the quotients in the boxes. The remaining letters will spell the answer to the riddle.

1. $5\overline{)95}$

2. $6\overline{)852}$

3. $4\overline{)940}$

4. $7\overline{)938}$

5. $3\overline{)762}$

6. $4\overline{)864}$

7. $3\overline{)744}$

8. $8\overline{)976}$

9. $5\overline{)965}$

| 485 A | 122 I | 19 T | | | | | | | | | |
|---|---|---|---|---|---|---|---|---|---|---|---|
| 216 S | 532 G | 121 I | 142 U | 322 R | 345 R | 235 P | 14 A | 18 F | 248 A | 238 I | 242 C |
| 193 B | 195 J | 134 E | 342 A | 254 T | 35 M | | | | | | |

"What do you call it when giraffes going in one direction get mixed-up with giraffes going in another direction?

___

___ ___ ___ ___ ___ ___ ___ ___   ___ ___ ___

# Extend Yourself with Division

Complete these division extensions.

15 ÷ 3 = ________

150 ÷ 3 = ________

1,500 ÷ 3 = ________

15,000 ÷ 3 = ________

28 ÷ 7 = ________

2,800 ÷ 7 = ________

28,000 ÷ 7 = ________

36 ÷ 6 = ________

360 ÷ 6 = ________

36,000 ÷ 6 = ________

480 ÷ 8 = ________

4,800 ÷ 8 = ________

48,000 ÷ 8 = ________

Write your own extensions to these division problems.

| | | | | |
|---|---|---|---|---|
| 18 ÷ 6 = ________ | 24 ÷ 6 = ________ |
| ________ ÷ 6 = ________ | ________ ÷ 6 = ________ |
| ________ ÷ 6 = ________ | ________ ÷ 6 = ________ |
| ________ ÷ 6 = ________ | ________ ÷ 6 = ________ |

Find the quotients for these division extensions.

54 ÷ 9 = ______

540 ÷ 90 = ______

5,400 ÷ 900 = ______

54,000 ÷ 9,000 = ______

42 ÷ 6 = ______

420 ÷ 60 = ______

4,200 ÷ 600 = ______

42,000 ÷ 6,000 = ______

32 ÷ 8 = ______

320 ÷ 80 = ______

3,200 ÷ 800 = ______

27 ÷ 9 = ______

270 ÷ 9 = ______

2,700 ÷ 90 = ______

27,000 ÷ 900 = ______

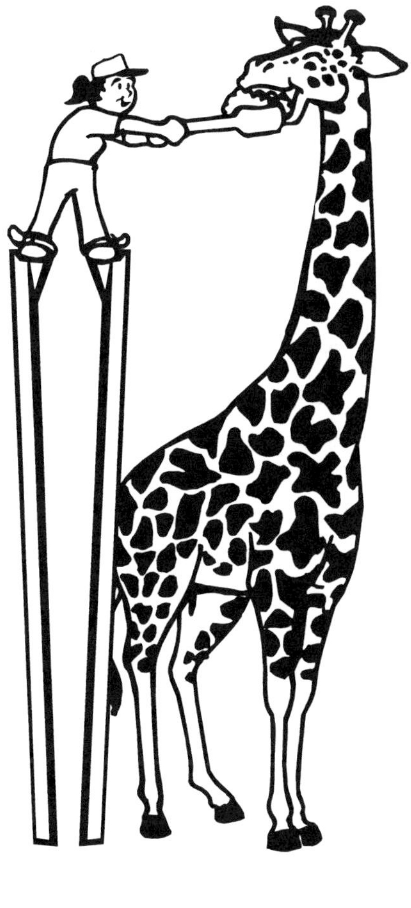

Make two sets of division extension problems of your own.

| | | | | | | | | | |
|---|---|---|---|---|---|---|---|---|---|
| ______ | ÷ | ______ | = | ______ | ______ | ÷ | ______ | = | ______ |
| ______ | ÷ | ______ | = | ______ | ______ | ÷ | ______ | = | ______ |
| ______ | ÷ | ______ | = | ______ | ______ | ÷ | ______ | = | ______ |
| ______ | ÷ | ______ | = | ______ | ______ | ÷ | ______ | = | ______ |

# Divide and Conquer

**Example:**

quotient → *15 R 3* ← remainder
divisor → $4\overline{)63}$ ← dividend

$$\begin{array}{r} 63 \\ \underline{4\phantom{0}} \\ 23 \\ \underline{20} \\ 3 \end{array}$$

Find the quotient.

1. $4\overline{)175}$
2. $7\overline{)253}$
3. $5\overline{)63}$
4. $3\overline{)89}$
5. $7\overline{)99}$
6. $8\overline{)93}$
7. $6\overline{)153}$
8. $9\overline{)128}$
9. $8\overline{)146}$

To check your work, multiply the quotient by the divisor. Then add the remainder.

**Example:**

```
   2 R 4        16
16 ) 36        x  2
     32        ----
     --          32
      4        +  4
               ----
                 36
```

1. $23\overline{)96}$

2. $16\overline{)86}$

3. $27\overline{)93}$

4. $13\overline{)88}$

5. $24\overline{)753}$

6. $48\overline{)687}$

 

# Determining Decimals

Match the decimals. Write the letters in the blanks below to answer the riddle. The first one is done for you.

| | | | |
|---|---|---|---|
| 1. | three tenths | 5.62 | U |
| 2. | two and four tenths | 0.03 | M |
| 3. | five and sixty-two hundredths | 0.003 | G |
| 4. | three hundredths | 0.3 | A |
| 5. | fifty-six and two tenths | 5.062 | N |
| 6. | two and four thousandths | 0.562 | R |
| 7. | five and sixty-two thousandths | 2.4 | H |
| 8. | three thousandths | 2.040 | E |
| 9. | two and forty thousandths | 2.004 | I |
| 10. | five hundred sixty-two thousandths | 56.2 | D |

**"What do you get when you cross a hummingbird with a doorbell?"**

| A | ___ | ___ | ___ | ___ | ___ | ___ | ___ | ___ | ___ |
|---|---|---|---|---|---|---|---|---|---|
| 1 | 2 | 3 | 4 | 5 | 6 | 7 | 8 | 9 | 10 |

Write these words in decimal form.

1. two and four tenths ____________
2. sixteen and twenty-seven hundredths ____________
3. eight and four hundredths ____________
4. twenty-six and seven tenths ____________
5. nineteen hundredths ____________

Write these decimals in words.

6. 5.3 ______________________________
7. 12.15 ______________________________
8. 1.06 ______________________________
9. 2.3 ______________________________
10. 3.03 ______________________________
11. 100.0 ______________________________
12. 12.77 ______________________________

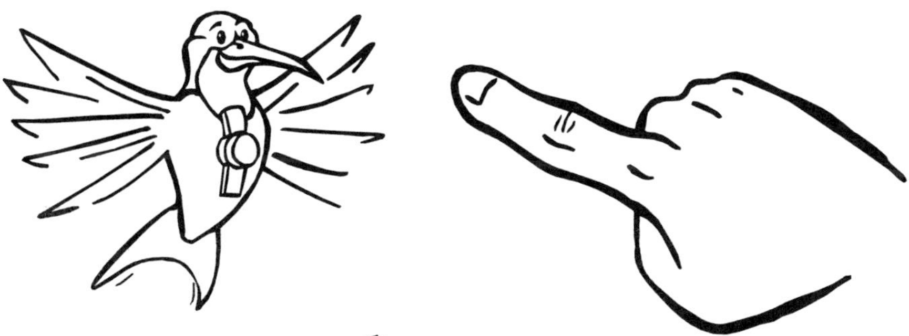

# How Do They Compare?

Circle the largest number in each row.

| | | | | | | |
|---|---|---|---|---|---|---|
| 1. | G | 0.031 | C | 0.13 | A | 0.8 |
| 2. | B | 1.01 | H | 1.10 | L | 1.010 |
| 3. | A | 2.301 | E | 2.056 | U | 2.30 |
| 4. | B | 0.709 | T | 0.76 | R | 0.09 |

Circle the smallest number in each row.

| | | | | | | |
|---|---|---|---|---|---|---|
| 5. | C | 0.001 | K | 0.01 | T | 0.100 |
| 6. | H | 5.03 | P | 5.1 | R | 5.031 |
| 7. | E | 9.10 | U | 9.70 | I | 9.07 |
| 8. | R | 7.001 | T | 7.0 | B | 7.20 |

**"What kind of tool does a baby bird need for getting out of an eggshell?"**

Write the circled letter from each problem above.

____ (1) ____ (2) ____ (3) ____ (4) ____ (5) ____ (6) - ____ (7) ____ (8)

Write <, >, or = to compare the decimals below. Circle the letter under the symbol used in each problem.

| | | | | < | > | = |
|---|---|---|---|---|---|---|
| 1. | 4.248 | ______ | 4.261 | C | T | R |
| 2. | 41.01 | ______ | 41.010 | O | U | R |
| 3. | 6.035 | ______ | 6.008 | K | O | P |
| 4. | 5.045 | ______ | 5.45 | A | K | B |
| 5. | 4.26 | ______ | 4.62 | K | O | W |
| 7. | 0.07 | ______ | 0.7 | A | M | C |
| 6. | 0.10 | ______ | 0.100 | I | E | C |
| 8. | 4.301 | ______ | 4.310 | O | V | S |
| 9. | 5.030 | ______ | 5.03 | L | R | L |
| 10. | 0.9 | ______ | 0.099 | A | A | L |

Write the circled letters from the chart above in order in the spaces below to answer the riddle.

**"What do frogs drink at picnics?"**

___ ___ ___ ___ ___ ___ ___ ___ ___ ___
1 2 3 4 5 6 7 8 9 10

# Order of the Day

List the numbers from least to greatest.

1. 3.62 3.32 3.091 ____________________
2. 5.3 5.6 5.5 ____________________
3. 71.20 71.38 71.3 ____________________
4. 3.01 3.10 3.011 ____________________
5. 1.7 1.723 1.72 ____________________

Fill in the missing numbers.

6. 0.07 0.08 0.09 ____ ____ 0.12 0.13 ____
7. 3.4 ____ 3.6 3.7 ____ 3.9 ____
8. 8.27 8.28 8.29 ____ ____ 8.32 8.33 ____
9. ____ ____ 4.1 4.2 4.3 ____ 4.5
10. 0.007 0.008 ________ 0.010 ________ 0.012

Order the numbers from least to greatest on the lines below. Then write the corresponding letters in the boxes.

| | | | | | |
|---|---|---|---|---|---|
| 3.6 | I | 3.00 | I | 2.1 | A |
| 4.01 | O | 3.61 | D | 4.45 | N |
| 2.111 | C | 4.09 | A | 2.9 | H |
| 4.06 | N | 3.999 | G | 3.9 | O |
| 3.4 | L | 4.2 | B | 4.4 | U |

**"What do you call a cold puppy that is sitting on a baby rabbit?"**

☐ ☐ ☐ ☐ ☐ ☐

____ ____ ____ ____ ____ ____

☐ ☐ ☐ ☐ ☐

____ ____ ____ ____ ____

☐ ☐ ☐ ☐

____ ____ ____ ____

# Round the Corner

Complete the chart by rounding each number as directed.

| | Whole Number | Tenth | Hundredth |
|---|---|---|---|
| 6.013 | | | |
| 23.278 | | | |
| 4.496 | | | |
| 8.021 | | | |
| 9.836 | | | |
| 4.997 | | | |
| 5.877 | | | |
| 7.391 | | | |
| 10.099 | | | |
| 100.990 | | | |

Round each decimal as directed. Cross out the boxes with the correct answers. The remaining letters will answer the riddle below.

Round . . .

1. 4.09 to the nearest tenth. ________
2. 24.31 to the nearest whole number. ________
3. 3.066 to the nearest hundredth. ________
4. 4.33 to the nearest tenth. ________
5. 5.88 to the nearest tenth. ________
6. 4.893 to the nearest hundredth. ________

| M | R | I | L | C | K | A | S |
|---|---|---|---|---|---|---|---|
| 3.08 | 4.1 | 4.90 | 4.4 | 24 | 4.0 | 4.89 | 4.0 |
| B | N | A | G | K | E | R | S |
| 5.9 | 25 | 3.1 | 3.07 | 4.09 | 5.8 | 4.3 | 5.6 |

**"What kind of reptiles do kids like to drink with their hamburgers?"**

____ ____ ____ ____

____ ____ ____ ____ ____ ____

# Sum Decimals

Find the sums. Write the corresponding letters in the riddle below.

1. 0.35 + 0.7 = ____ L
2. 3.816 + 2.719 = ____ U
3. 7.1 + 32.90 = ____ O
4. 6.831 + 4.9 = ____ I
5. 5.7 + 2.368 = ____ T
6. 2.010 + 99.99 = ____ M
7. 0.213 + 0.937 = ____ S
8. 3.446 + 1.71 = ____ P
9. 1.8 + 0.09 = ____ F

**"What kind of flowers grow under your nose?"**

| ___ | ___ | ___ | ___ | ___ | ___ |
|---|---|---|---|---|---|
| 8.068 | 6.535 | 1.05 | 11.731 | 5.156 | 1.150 |

Find the sum. Cross out the answers in the boxes. The remaining letters will answer the ridddle.

1. 3.41 + 2.601 = __________
2. 4.63 + 1.033 = __________
3. 3.8 + 2.012 = __________
4. 2.61 + 2.7 = __________
5. 3.6 + 0.47 = __________
6. 0.011 + 3.67 = __________

| | | | | | | |
|---|---|---|---|---|---|---|
| M 3.40 | B 3.681 | I 3.13 | N 5.41 | R 6.011 | U 6.1 | C 5.663 |
| T 7.2 | E 8.9 | G 5.31 | M 3.99 | I 1.71 | K 4.07 | C 2.76 |
| Y 5.812 | E 3.00 | | | | | |

**"What do cats eat when they are in a hurry?"**

____ ____ ____ ____ ____ ____

____ ____ ____ ____

# Take It Away

Find the differences.

1. $\begin{array}{r} 61.32 \\ -\ 42.3 \\ \hline \end{array}$

2. $\begin{array}{r} 4.813 \\ -\ 2.32 \\ \hline \end{array}$

3. $\begin{array}{r} 5.8 \\ -\ 4.13 \\ \hline \end{array}$

4. $\begin{array}{r} 14.24 \\ -\ 1.031 \\ \hline \end{array}$

5. $\begin{array}{r} 54.27 \\ -\ 3.36 \\ \hline \end{array}$

6. $\begin{array}{r} 0.581 \\ -\ 0.19 \\ \hline \end{array}$

7. 14.3 – 3.21 = ________

8. 6.53 – 1.3 = ________

Find the sum or difference. Show your work for the number stories.

1. $31.47
   16.21
   + 18.37

2. $47.03
   − 12.85

3. $1.35 + $12.23 = ________

4. $16.27 + $8.15 = ________

5. Last summer Chris earned $83.75 mowing lawns, $23.50 washing cars, and $68.45 babysitting. How much did he earn from all three jobs?

6. How much more did Chris earn from mowing lawns than from babysitting?

# Land and Sea

Find the products. Use the corresponding letters to solve the riddle below.

| | | |
|---|---|---|
| 1. 3.61 x 5 = ____ H | 2. 4.27 x 8 = ____ R | 3. 6.42 x 3 = ____ I |
| 4. 2.31 x 21 = ____ C | 5. 3.70 x 43 = ____ I | 6. 0.081 x 36 = ____ L |
| 7. 6.17 x 314 = ____ S | 8. 8.21 x 476 = ____ O | 9. 6.322 x 57 = ____ G |

**"What do you get when you bury money in your garden?"**

| ____ | ____ | ____ | ____ |
|---|---|---|---|
| 34.16 | 159.10 | 48.51 | 18.05 |

| ____ | ____ | ____ | ____ |
|---|---|---|---|
| 1,937.38 | 3,907.96 | 19.26 | 2.916 |

Find the quotients. Show your work. Then use the corresponding letters to solve the riddle below.

**1.** $6\overline{)25.92}$ **S**

**2.** $4\overline{)94.8}$ **W**

**3.** $5\overline{)1.875}$ **A**

**4.** $7\overline{)37.38}$ **E**

**5.** $3\overline{)1.449}$ **S**

**6.** $8\overline{)280.8}$ **A**

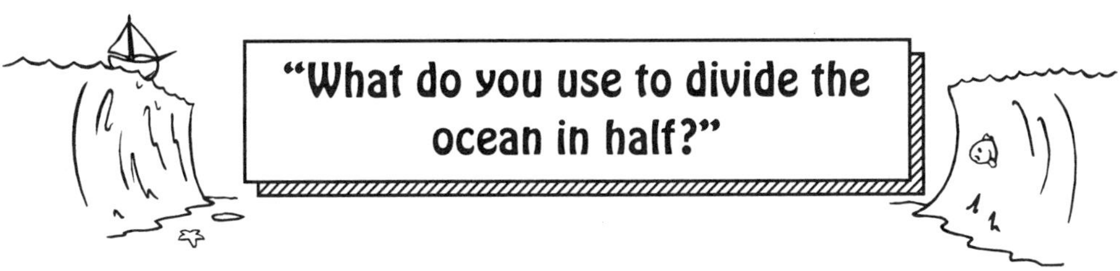

**"What do you use to divide the ocean in half?"**

**A** ___ ___ ___ – ___ ___ ___

0.483 5.34 0.375 – 4.32 35.1 23.7

# Zero Zonkers

Find the products. Use the corresponding letters to solve the riddle below.

1. 13.4 x 0.2 = ____ M

2. 8.7 x 5 = ____ N

3. 1.63 x 0.4 = ____ E

4. 3.4 x 6 = ____ A

5. 214 x 0.3 = ____ L

6. 9.9 x 4 = ____ O

7. 2.3 x 0.4 = ____ I

8. 3.31 x 4 = ____ S

9. 10.1 x 0.5 = ____ G

10. 4.6 x 3 = ____ T

**"What did the little crab say when the tide finally came in?"**

___ ___ ___ ___    ___ ___ ___ ___
64.2 39.6 43.5 5.05    13.8 0.92 2.68 0.652

___ ___    ___ ___ ___
43.5 39.6    13.24 0.652 20.4

Find the product. Use the corresponding letters to solve the riddle below.

1. $\begin{array}{r} 0.76 \\ \times\ 1.3 \\ \hline \end{array}$ U

2. $\begin{array}{r} 6.13 \\ \times 0.05 \\ \hline \end{array}$ L

3. $\begin{array}{r} 24.6 \\ \times\ 2.3 \\ \hline \end{array}$ E

4. $\begin{array}{r} 68.1 \\ \times 0.04 \\ \hline \end{array}$ A

5. $\begin{array}{r} 34.3 \\ \times 0.14 \\ \hline \end{array}$ P

6. $\begin{array}{r} 7.8 \\ \times 0.03 \\ \hline \end{array}$ D

**"What do you call a snowman on a warm day?"**

____    ____ ____ ____ ____ ____ ____

2.724    4.802 0.988 0.234 0.234 0.3065 56.58

# Take Me Out to the Ballgame

| Souvenir Stand | |
|---|---|
| Autographed Baseballs | $24.95 |
| Baseball Hats | $12.75 |
| Programs | $2.50 |
| Pennants | $4.75 |
| T-shirts | $13.49 |
| Sweatshirts | $22.25 |
| Batting Helmets | $4.50 |
| Water Bottle | $2.75 |
| Mini Poster | $3.50 |
| Life Size Poster | $9.50 |
| Coloring Book | $2.25 |

Use the chart on page 68 to solve these money problems. Remember to show your work and label the answers.

1. Kyle wants to buy a mini-poster for each of his three little sisters. How much will it cost for 3 posters? Kyle has a $20 bill. How much change will Kyle receive?

2. Mr. Matthews has $50 to spend on prizes for the school candy sale. He wants to buy an autographed baseball, a helmet, a T-shirt, and a pennant. How much will he spend?

   With the change from his $50, what else can Mr. Matthews buy from the souvenir stand?

3. The souvenir stand sold 42 sweatshirts on a cold day at the ballpark. How much money did they take in from the sale of the sweatshirts?

   The souvenir stand bought the sweatshirts for $12 each. How much profit did they make on each sweatshirt? How much profit did they make on the 42 sweatshirts?

# Chart It

The Eagles played 6 basketball games. Find the total points scored per game and the season average for each player.

| | Game in Series | | | | | | |
|---|---|---|---|---|---|---|---|
| **Player** | 1 | 2 | 3 | 4 | 5 | 6 | Player Averages |
| Laura | 2 | 3 | 4 | 2 | 3 | 4 | |
| Mandy | 5 | 4 | 4 | 6 | 8 | 9 | |
| Jan | 12 | 10 | 13 | 6 | 7 | 12 | |
| Nancy | 6 | 12 | 8 | 5 | 3 | 8 | |
| Mary | 2 | 2 | 10 | 12 | 10 | 0 | |
| Amy | 10 | 4 | 3 | 8 | 4 | 7 | |
| Carol | 5 | 2 | 1 | 3 | 0 | 7 | |
| Game Totals | | | | | | | |

Use each rule to solve the problems and complete the charts.

| N | Rule: N + 5 |
|---|---|
| 3 | 3 + 5 = 8 |
| 6 | |
| 9 | |
| 12 | |
| 16 | |

| N | Rule: N x 4 |
|---|---|
| 3 | |
| 5 | |
| 7 | |
| 8 | |
| 9 | |

| N | Rule: (n + 2) x 3 |
|---|---|
| 6 | |
| 8 | |
| 10 | |
| 5 | |
| 4 | |

| N | Rule: (n + 3) ÷ 4 |
|---|---|
| 17 | |
| 1 | |
| 13 | |
| 33 | |
| 5 | |

# Mystery Number

Here is a little trick that you can do with your house number and your age.

First try it with your own house number and age. Then try it out on a friend.

Write your house number. __________

Multiply by 2. __________

Add 5. __________

Multiply by 50. __________

Add your age in years. __________

Add 365. __________

Subtract 615. __________

Can you find your house number and age in the answer?